ISNM 69:
International Series of Numerical Mathematics
Internationale Schriftenreihe zur Numerischen Mathematik
Série internationale d'Analyse numérique
Vol. 69

Edited by
Ch. Blanc, Lausanne; R. Glowinski, Paris;
G. Golub, Stanford; P. Henrici, Zürich;
H.O. Kreiss, Pasadena; A. Ostrowski, Montagnola;
J. Todd, Pasadena

Birkhäuser Verlag
Basel · Boston · Stuttgart

Numerical Treatment of Eigenvalue Problems
Vol. 3

Workshop in Oberwolfach, June 12–18, 1983

Numerische Behandlung von Eigenwertaufgaben
Band 3

Tagung in Oberwolfach, 12.–18. Juni 1983

Edited by
Herausgegeben von

J. Albrecht
L. Collatz
W. Velte

1984

Birkhäuser Verlag
Basel · Boston · Stuttgart

Editors/Herausgeber

J. Albrecht
Technische Universität
Clausthal
Institut für Mathematik
D–3392 Clausthal-Zellerfeld (FRG)

L. Collatz
Universität Hamburg
Institut für
Angewandte Mathematik
Bundesstrasse 55
D–2 Hamburg 13 (FRG)

W. Velte
Institut für Angewandte
Mathematik und Statistik
der Universität
Am Hubland
D–8700 Würzburg

CIP-Kurztitelaufnahme der Deutschen Bibliothek

Numerical treatment of eigenvalue problems:
workshop . . . = Numerische Behandlung von
Eigenwertaufgaben. – Basel; Boston; Stuttgart:
Birkhäuser
  Bis Bd. 2 (1979) u.d.T.: Numerische Behandlung
  von Eigenwertaufgaben

Vol. 3. Workshop in Oberwolfach, June 12 – 18,
1983. – 1984.
  (International series of numerical mathematics;
  Vol. 69)
  ISBN 3-7643-1605-5

  NE: PT; GT

© 1984 Birkhäuser Verlag Basel
Printed in Switzerland by Birkhäuser AG, Graphische Unternehmen, Basel
ISBN 3-7643-1605-5

PREFACE

This volume contains the manuscripts of talks given at a conference on
"Numerical Treatment of Eigenvalue Problems" held in June 1983 at the
Mathematical Research Institute, Oberwolfach.

The meeting centered in part of methods for the computation of eigenvalue
bounds with reports on new developments of various methods (method of Lehmann-
Maehly, methods of intermediate problems, estimates using difference methods,
inclusion by quotients theorems); numerous examples arising out of physics and
engineering sciences (computation of rolling frequences, eigenfrequences of
plates and membranes, buckling values, energy levels of atoms and critical
values in flow problems) helped to illustrate the significance of these results.
Some emphasis was also laid on matrix eigenvalue problems; next to theoretical
results (e.g. perturbation theorems), numerical algorithms for the treatment
of such problems were discussed, in particular in the form arising from the
use of finite elements. There were also reports on problems with nonlinear
operators occuring in the treatment of nonlinear oscillations (Duffing differential equation) and in the physics of plasma.

As always in the stimulating atmosphere of the Oberwolfach institute, extensive
discussions contributed to intensify the flow of ideas between the participants,
among whom a number of guests from foreign European countries and America.

The hearty thanks of the participants go to the director of the Mathematical
Research Institute, Herrn Prof. Dr. M. Barner, who, at a time of increasing
importance of numerical and applied mathematics, made it possible to held this
meeting; to his collaborators in Oberwolfach and Freiburg; and to Birkhäuser
Verlag for the excellent presentation of the volume.

Julius Albrecht	Lothar Collatz	Waldemar Velte
Clausthal-Zellerfeld	Hamburg	Würzburg

6

V O R W O R T

Der Band enthält Manuskripte zu Vorträgen, die auf einer Tagung am Mathemati-
schen Forschungsinstitut Oberwolfach über "Numerische Behandlung von Eigenwert-
aufgaben" im Juni 1983 gehalten wurden.
Einer der Schwerpunkte der Tagung lag bei den Verfahren zur Berechnung von
Eigenwertschranken. Berichtet wurde über Weiterentwicklungen verschiedener
Methoden (Lehmann-Maehly-Verfahren, Methoden der "intermediate problems",
Abschätzungen mit Hilfe von Differenzenverfahren, Quotienten-Einschliessungs-
sätze); die Bedeutung der hierbei erzielten Resultate wurde an Hand zahlreicher
Beispiele aus der Physik und den Ingenieurwissenschaften (Berechnung von
Schlingerfrequenzen, von Schwingungsfrequenzen von Platten und Membranen, von
Beulwerten, von Energieniveaus von Atomen und von kritischen Werten bei
Strömungsproblemen) erläutert.
Ein weiterer Schwerpunkt lag bei den Eigenwertaufgaben mit Matrizen; hier
wurden neben theoretischen Resultaten (z.B. Störungssätzen) insbesondere
numerische Algorithmen zur Behandlung solcher Matrix-Eigenwertaufgaben, wie sie
bei der Verwendung der Methode der finiten Elemente auftreten, besprochen.
Ferner wurde über einige Aufgaben mit nichtlinearen Operatoren, die bei der
Behandlung nichtlinearer Schwingungen (Duffingsche Differentialgleichung) und
in der Plasmaphysik vorkommen, berichtet.
Wie immer trugen in der anregenden Atmossphäre des Oberwolfacher Instituts auch
ausführliche Diskussionen wesentlich dazu bei, den Gedankenaustausch zwischen
den Tagungsteilnehmern, unter denen sich zahlreiche Gäste aus dem europäischen
Ausland und aus Amerika befanden, zu intensivieren.
Der herzliche Dank der Teilnehmer gilt dem Direktor des Mathematischen
Forschungsinstitut, Herrn Professor Dr. M. Barner, der es in einer Zeit
wachsender Bedeutung der Numerischen und Angewandten Mathematik ermöglichte,
diese Tagung durchzuführen, sowie seinen Mitarbeitern in Oberwolfach und
Freiburg, und dem Birkhäuser Verlag für die wie stets sehr gute Ausstattung
des Bandes.

Julius Albrecht   Lothar Collatz   Waldemar Velte
Clausthal-Zellerfeld   Hamburg   Würzburg

I N D E X

ISNM. Vol. 69
Numerical Treatment of
Eigenvalue Problems, Vol. 3
© 1983 Birkhäuser Verlag Basel

THE COMPUTATION OF CONVERGENT LOWER BOUNDS
IN QUANTUM MECHANICAL EIGENVALUE PROBLEMS

Christopher Beattie

A computational approach guaranteeing convergent
lower bounds to eigenvalues of a wide class of Schrödinger
operators is introduced. After motivating the necessary
strategy in the simple two-particle case, a variant of Fox's
intermediate Hamiltonian construction is presented for complex
atoms, together with the necessary convergence criteria.
Computational features of Fox's method are discussed in the
context of sparse matrix strategies that can utilize the
algebraic structure of the resulting matrices to advantage.

§1. <u>Introduction</u>. An important theoretical and practical
problem in quantum mechanics is the computation of Schrödinger
operator eigenvalues to very high accuracy. Schrödinger
operators govern the energy of atoms and atomic ions in
accordance with conjectured physical principles describing the
interactions among the constitutive particles. Eigenvalues of
such operators are related to atomic energy levels, which may
be determined quite accurately from experimental spectroscopic
data. Comparison with computed eigenvalues for the modeling
operator then provides a measure of how accurately the
operator models the actual atomic system. The computed
eigenvalues themselves are useful in determining whether
electronic shell configurations exist and are in fact
determined by the operator. Unfortunately it is rarely
possible to find exact values for the operator eigenvalues
explicitly, so it is necessary to consider computational
schemes for obtaining accurate estimates instead.

The problem of assessing the accuracy of an estimate
to any given eigenvalue is evidently equivalent to the problem
of obtaining rigorous upper and lower bounds to that
eigenvalue. Computational procedures that provide rigorous
upper bounds are well known and highly refined at this point,
e.g. Rayleigh-Ritz (configuration interaction) and Hartree-
Fock methods. In contrast, lower bound methods are yet in
somewhat of a germinal state. Even so, sufficient knowledge
and computational experience has been gained in the last
twenty years to indicate a much more complex situation
requiring much more subtle analysis and somewhat more
computational effort than appears to be required in
complimentary upper bound methods.

§2. The Class of Problems to be Treated. We consider the
task of estimating the bound state energies of atomic systems
having m identical particles interacting with each other and
with an infinitely massive nucleus. The model Hamiltonian is
taken as

$$H = \sum_{k=1}^{m} [\frac{-1}{2\mu} \Delta_k + W(\underset{\sim}{r}_k)] + \sum_{i<j}^{m} V(\underset{\sim}{r}_i - \underset{\sim}{r}_j)$$

operating on a domain dense in $L^2(\mathbb{R}^{3m})$. In this
notation $\{\Delta_k\}$ are Laplacians acting in copies of $L^2(\mathbb{R}^3)$;
$\{\underset{\sim}{r}_k\}$ are position vectors in $\mathbb{R}^3$; W and V are
potentials on $\mathbb{R}^3$ representing nuclear and interparticle
forces, respectively; and μ is the particle mass. Generally
W and V are simple Coulombic interactions, however an
interesting alternative was offered by Thirring [17], and
explored extensively by Greenlee and Russell ([12], [16]).
There, W is taken as an effective screening potential so
that V becomes the residual correction needed to regain the
exact Coulombic energies.

In order to construct convergent lower estimates to the isolated eigenvalues of H , we require the following three assumptions:

1)   The potentials V and W are Kato potentials, i.e. members of $L^2(\mathbb{R}^3) + [L^\infty(\mathbb{R}^3)]_\varepsilon$,

2) $V \geqslant 0$ a.e. in $\mathbb{R}^3$, and

3)   the self-adjoint operator corresponding to $\frac{-1}{2\mu} \Delta + W$ in $L^2(\mathbb{R}^3)$ should have discrete eigenvalues below the lowest point of the essential spectrum that are computable to arbitrarily high precision together with their corresponding eigenfunctions.

Some observations may be made immediately with regard to these assumptions. We first note that the self-adjoint operator associated with

$$H_0 = \sum_{k=1}^{m} [\frac{-1}{2\mu} \Delta_k + W(\underline{r}_k)]$$

has discrete lower spectra that is computationally resolvable in terms of the spectra of $\frac{-1}{2\mu} \Delta + W$ by separation of variables. Furthermore, the symmetric operator

$$\hat{H} = \sum_{i<j}^{m} V(\underline{r}_i - \underline{r}_j)$$

is essentially self-adjoint on a common core with both H_0 and H . Since $\hat{H} \geqslant 0$, we see that $H_0 \leqslant H$ (in the usual sense of ordering for symmetric operators, cf. [18]). Hence we may conclude from standard monotone theorems ([18]) that the lower eigenvalues of H_0 are lower bounds to the corresponding eigenvalues of H and that the lowest point of the essential spectrum of H_0 is a lower bound to the lowest point of the essential spectrum of H . The lower bounds provided by H_0

are generally very poor, so we are led naturally to seek out ways to improve them. This is commonly done by approximating the positive definite operator $\hat{H}$.

In 1950 Aronszajn [1] presented in a broader context a systematic method of approximating $\hat{H}$ from below by constructing a bounded positive semi-definite operator $0 < \hat{H}^* < \hat{H}$. The monotone theorems appealed to above then indicate that the eigenvalues of $H^* = H_0 + \hat{H}^*$ lie intermediate between those of H_0 and those of H, i.e. they are improved lower bounds to the eigenvalues of H.

The construction begins with a family of vectors $\{p_\nu\}$ chosen in $\mathrm{Dom}(\hat{H})$. An associated family of projections $\{P^\beta\}$ may then be defined on $L^2(\mathbb{R}^3)$ as

$$P^\beta u = \sum_{k,\ell=1}^{\beta} (u, \hat{H}p_k) b_{k\ell} p_\ell$$

where $(\cdot, \cdot)$ denotes the inner product on $L^2(\mathbb{R}^{3m})$ and $b_{k\ell}$ are the elements of the matrix inverse to the weighted Gram matrix $[(p_i, \hat{H}p_j)]$. One may check directly that $\hat{H}P^\beta$ is a bounded, positive semi-definite operator on $L^2(\mathbb{R}^{3m})$ and that $0 < \hat{H}P^{\beta_1} < \hat{H}P^{\beta_2} < \hat{H}$ for $\beta_1 < \beta_2$ as a consequence of Bessel's inequality in the $\hat{H}$ - weighted inner product space associated with $\mathrm{Dom}(\hat{H})$. The operators defined by $H^\beta = H_0 + \hat{H}P^\beta$ satisfy $H_0 < H^{\beta_1} < H^{\beta_2} < H$ for $\beta_1 < \beta_2$ hence the eigenvalues of H^β are nondecreasing in β and always lower bounds to the corresponding eigenvalues of H (thus they are "improvable" lower bounds).

§3. <u>Bazley's Special Choice and Convergence</u>. Aronszajn showed that these improvable lower bounds are computable in principle from information available for H_0, however in practical terms, the required computation is generally infeasible. In 1959, Bazley [2] discovered that for a special choice of $\{p_\nu\}$, the spectrum of H^β may be explicitly computed through the diagonalization of a symmetric matrix of order β. In fact, if $\{u_i^0\}$ denote the eigenvectors of H_0 and $\{u_i^0\} \subset \text{Ran}(\hat{H})$, one may choose $p_k = \hat{H}^{-1}u_k^0$. It is then quite easy to see that $U_\beta = \text{span}_\beta\{u_k^0\}$ is a reducing space for H^β. Restricted to U_β, H^β is equivalent to a matrix operator on $\mathbb{R}^\beta$, whereas restricted to $U_\beta^\perp$, H^β is equivalent to H_0.

Although Bazley's special choice produces a tractable computational problem, this choice of projecting vectors is poor from the viewpoint of convergence. To clarify this point it is useful to review what is currently known regarding the convergence of Aronszajn's method in the presence of essential spectrum.

One may immediately notice that $\text{Ran}(\hat{H}P^\beta)$ is finite-dimensional hence $\hat{H}P^\beta$ is a compact operator on $L^2(\mathbb{R}^{3m})$, implying in turn that $\sigma_{ess}(H^\beta) = \sigma_{ess}(H_0)$ (Weyl's Theorem). Thus the <u>only</u> eigenvalues of H that are accessible to convergent estimates are those that lie <u>below</u> $\inf \sigma_{ess}(H_0)$ (the lowest point in the essential spectrum of H_0).

With this proviso implicit, we may state convergence criteria due to Brown [6] and Greenlee [11] (a third distinct proof may be found in [4]): If $\hat{H}^{-1}$ extends to a bounded operator then it suffices to require that span $\{p_\nu\}$ is dense in $\text{Dom}(\hat{H})$ endowed with the graph norm $(||u||^2 + ||\hat{H}u||^2)^{1/2}$ in order to guarantee convergence of the eigenvalues of H^β to the corresponding accessible eigenvalues of H.

This indicates the sort of result we seek, however $\hat{H}$ will not generally have a bounded inverse or perhaps not even dense range, hence the following criteria is of more

use ([4]): If $\hat{H}$ is essentially self-adjoint on a core for
H then it is sufficient to require density
of $\{\hat{H}p_\nu\}$ in $\overline{\text{Ran}}\ (\hat{H})$ with respect to the L^2-norm.

Since the eigenfunctions of H_0 are not dense in
L^2 whenever $\sigma_{ess}(H_0)$ contains a line segment (which includes
our case), Bazley's special choice does not satisfy this
convergence criteria. In fact, a stronger statement may be
made. If an eigenvector of H does not lie in
$\overline{\text{span}}\ \{u_k^0\}$, the special choice estimates to the corresponding
eigenvalue __will__ __not__ converge to the proper value.

Although the method of special choice has frequent
computational advantages, we do not consider it as a reliable
method for problems containing significant essential
spectrum. We must relax the constraints imposed by
computational feasibility so as to allow a choice of
projecting vectors $\{p_\nu\}$ that does satisfy the convergence
conditions given.

§__4__. __Truncation Including the Remainder__. Bazley and Fox [3]
noticed that a computationally feasible lower bound problem
can be constructed with arbitrary $\{p_\nu\} \subset \text{Dom}\ (\hat{H})$ if a slightly
inferior base operator is used. The new base operator is
defined in terms of H_0 as a spectral truncation:

$$H_0^n u = \sum_{k=1}^{n} \mu_k^0 (u,\ u_k^0) u_k^0 + \mu_{n+1}^0 [u - \sum_{k=1}^{n} (u,\ u_k^0) u_k^0],$$

where $\{\mu_k^0, u_k^0\}$ are eigenvalue-eigenvector pairs for H_0.
H_0^n is equivalent to H_0 on $\mathcal{U}_n$ and to $\mu_{n+1}^0 I$ on $\mathcal{U}_n^\perp$.
Thus H_0^n is a bounded self-adjoint operator
and $H_0^n < H_0 < H$. The interesting feature of H_0^n is that it is a
translation by $\mu_{n+1}^0 I$ of a finite-rank operator with range
$\mathcal{U}_n$, hence __any__ subspace containing $\mathcal{U}_n$ is a reducing space

for H_0^n. If we consider intermediate problems defined by

$$H^{n\beta} = H_0^n + \hat{H}P^\beta \prec H_0 + \hat{H}P^\beta \prec H,$$

we may obtain improvable lower bounds to the eigenvalues of
H by resolving $H^{n\beta}$ on the $(n + \beta)$-dimensional
subspace $u_n \oplus \operatorname{span}_\beta \{\hat{H}p_\nu\}$ which reduces $H^{n\beta}$ for any choice
of $\{p_\nu\}$. On the orthogonal complement of this subspace we
have $H^{n\beta} = \mu_{n+1}^0 I$. The projecting family $\{p_\nu\}$ may now be
chosen to satisfy any density requirement desired, however due
to the spectral information for H_0 that was lost in
truncation, convergence of estimates to an eigenvalue of H
<u>will</u> <u>not</u> occur if the corresponding eigenvector of H does
not lie entirely within $\overline{\operatorname{span}}\,\{u_k^0\}$ (the same necessary
condition as seen for special choice). We may recover this
spectral information in part by approximating the truncation
remainder $\tilde{H}_0^n = H_0 - H_0^n$ in a way analogous to the way $\hat{H}$ was
approximated: choose $\{q_k\} \subset \operatorname{Dom}(H_0) \backslash u_n$ and define

$$Q^\alpha u = \sum_{i,j=1}^\alpha (u,\, \tilde{H}_0^n q_i) c_{ij} q_j$$

with $[c_{ij}]^{-1} = [(q_k,\, \tilde{H}_0^n q_\ell)]$. Then the intermediate operators

$$H^{n\alpha\beta} = H_0^n + \tilde{H}_0^n Q^\alpha + \hat{H}P^\beta$$

provide improvable lower bounds to the eigenvalues of H.
They are reduced by the $(n + \alpha + \beta)$-dimensional subspaces
$u_n \oplus \operatorname{span}_\alpha (\tilde{H}_0^n q_k) \oplus \operatorname{span}_\beta (\hat{H}p_k)$ (on the orthogonal complement,
$H^{n\alpha\beta} = \mu_{n+1}^0 I$, as before). Aside from the evaluation of inner
products, the computational effort involved consists
principally of the diagonalization of a matrix of
order $n + \alpha + \beta$. Moreover, the projecting

families $\{q_k\}$ and $\{p_k\}$ may be chosen to satisfy density conditions sufficient to guarantee convergence of every estimate to the appropriate accessible eigenvalue of H.

A simple example of an appropriate choice of projecting families will be given for estimating helium bound states. The Hamiltonian we consider is given in atomic units by

$$H = [- \frac{1}{2} \Delta_1 - \frac{1}{2} \Delta_2 - \frac{2}{|\underset{\sim}{r}_1|} - \frac{2}{|\underset{\sim}{r}_2|}] + \frac{1}{|\underset{\sim}{r}_1 - \underset{\sim}{r}_2|} \quad \text{on} \quad L^2(\mathbb{R}^6)$$

The bracketed term constitutes the hydrogen-like operator H_0 and the last term becomes $\hat{H}$. Since $\hat{H}$ is relatively compact with respect to H_0, $\sigma_{ess}(H_0) = \sigma_{ess}(H)$. Thus every lower eigenvalue of H is accessible to convergent estimates.

In order to construct convergent intermediate Hamiltonians as described above, we must select $\{p_\nu\}$ and $\{q_\nu\}$. Introduce families $\{\phi_k\}$ and $\{\psi_k\}$ in $L^2(\mathbb{R}^3)$ so that each ψ_k has exponential decay at ∞, $\{\psi_k\}$ is dense in $L^2(\mathbb{R}^3)$, and $\{\phi_k\}$ is dense in $W^{2,2}(\mathbb{R}^3)$. For example, we may take $\{\psi_k\}$ as the complete set of Laguerre polynomials and $\{\phi_k\}$ as the set of isotropic Slater orbitals [12]. Define

$$p_{\nu_{k\ell}} = |\underset{\sim}{r}_1 - \underset{\sim}{r}_2| \, \psi_k(\underset{\sim}{r}_1) \, \psi_\ell(\underset{\sim}{r}_2)$$

and

$$q_{\nu_{k\ell}} = \phi_k(\underset{\sim}{r}_1)\phi_\ell(\underset{\sim}{r}_2)$$

The resulting families $\{p_\nu\}$ and $\{q_\nu\}$ fulfill the appropriate density conditions given previously hence the corresponding intermediate Hamiltonians provide convergent estimates. The exponential decay condition is included only to assure that each $p_\nu \in L^2(\mathbb{R}^6)$. Polynomial decay of sufficiently high order

REFERENCES

1. Baker,C.T.H., The Numerical Treatment of Integral Equations,
   Clarendon Press Oxford (1977)
2. Baker,C.T.H., Miller,G.F., Treatment of Integral Equations by
   Numerical Methods, Academic Press London New York (1982)
3. Bulirsch,R., Stoer,J., Numerical Treatment of Ordinary Differential
   Equations by Extrapolation Methods, Num.Math.8,1-13 (1966)
4. Deuflhard,P., Bauer,H.J., A Note on Romberg Quadrature,
   preprint no. 169 of the SFB 123 Heidelberg (1982)
5. Gear,C.W., Numerical Initial Value Problems in Ordinary Differential
   Equations,Prentice Hall (1971)
6. Hairer,E., Lubich,Ch., Nørsett,S.P., Order of Convergence of One-Step
   Methods for Volterra Integral Equations of the second kind,
   preprint no. 137 of the SFB 123 Heidelberg (1981)
7. Hock,W., An Extrapolation Method with Step Size Control for Nonlinear
   Volterra Integral Equations, Num.Math.38,155-178 (1981)
8. Shampine,L.F., Gordon,M.K., Computer Solution of Ordinary Differential
   Equations, Freeman and Company (1975)

Dr. Herbert Arndt, Institut für Angewandte Mathematik der Universität
Bonn, Wegelerstr. 6, D-5300 Bonn

ISNM, Vol. 67
Numerical Methods of
Approximation Theory, Vol. 7
© 1983 Birkhäuser Verlag Basel

RESTABSCHÄTZUNGEN ZUR POLYNOMAPPROXIMATION

Helmut Braß

We prove $E_n[f] < \text{const}_k \, (\sqrt{n})^{k+1} \, 2^{-n}(n+1)!^{-1} \| f^{(n-k)} \|$ (see (1), (2) for the relevant definitions). This bound is optimal with respect to the order. The degree of approximation is realized by interpolation with Chebyshev knots and by expansion in terms of Chebyshev polynomials.

1. Einleitung

Es sei

$$(1) \qquad \| f \| = \sup_{-1 \le x \le 1} | f(x) |$$

$$(2) \qquad E_n[f] = \inf_{p \in P_n} \| f - p \| ,$$

wobei P_n die Menge der Polynome n-ten Grades bezeichnet. Ein Hauptproblem der quantitativen Theorie der Polynomapproximation ist die Bestimmung der besten Konstanten $\rho_{n,1}$ in der Abschätzung

$$E_n[f] \le \rho_{n,1} \| f^{(1)} \| , \quad 1=1,2,\ldots,n+1$$

$$n=1,2,\ldots .$$

Anders als im trigonometrischen Fall ist hier eine vollständige Lösung noch nicht erreicht. Sinwel [10] hat Abschätzungen angegeben, die für festes 1 und $n\rightarrow\infty$ asymptotisch scharf sind; für kleine n und 1 hat Ehresmann [6] Schranken numerisch bestimmt (die übrigens für $2 \leq 1 \leq n$ nicht scharf sind); der Fall, daß n und 1 beide groß sind, in dem Sinwels Schranken zu einer Überschätzung führen, soll hier behandelt werden.

Dazu werden Zahlen c_k definiert durch

$$c_k = \int_{-\infty}^{\infty} e^{-x^2} |H_k(x)| \, dx \ ,$$

wobei H_k das Hermite-Polynom

$$H_k(x) = (-1)^k \, e^{x^2} \, D^k [e^{-x^2}], \qquad (D[f]:=f')$$

bedeutet. Hiermit hat man

<u>Satz 1</u>   Sei $k\in\{0,1,\ldots\}$ fest. Für $n\rightarrow\infty$ gilt

$$\rho_{n,n-k} = \frac{c_{k+1}}{\sqrt{\pi}} \ \frac{(\sqrt{n})^{k+1}}{2^n(n+1)!} \ (1+o(1)).$$

Für die meisten Anwendungen genügt die folgende schwächere Form:

<u>Folgerung</u>   Zu jedem k gibt es eine Zahl C_k mit

$$E_n[f] \leq C_k \ \frac{(\sqrt{n})^{k+1}}{2^n(n+1)!} \ \| f^{(n-k)} \| \ .$$

In der praktischen Mathematik ist von Belang, daß die genannte Approximationsgüte mit sehr einfach zu konstruierenden Polynomen erreicht werden kann. Zur Präzisierung dieser Aussage definieren wir als "Standardverfahren" der Polynomapproximation die folgenden Operatoren $S_n^{(i)}=S_n$, die $C[-1,1]$ in P_n abbilden:

$$S_n[f] = \frac{a_0}{2} + \sum_{\nu=1}^{n} a_\nu T_\nu \quad , \quad a_\nu = A_i[fT_\nu]$$

$$(T_n \text{ die Tschebyscheff-Polynome})$$

mit einem der folgenden Funktionale

$$A_1[f] = \frac{2}{\pi} \int_{-1}^{1} f(x)dx$$

$$A_2[f] = \frac{2}{m} \sum_{\kappa=1}^{m} f\left(\cos \frac{2\kappa-1}{2m} \pi\right) \qquad m \geq n+1$$

$$A_3[f] = \frac{2}{m} \sum_{\kappa=0}^{m} {}'' f\left(\cos \frac{\kappa\pi}{m}\right) \qquad m \geq n+1 \; .$$

$S_n^{(1)}$ ist die Teilsumme der Tschebyscheff-Entwicklung. $S_n^{(2)}$ mit
$m=n+1$ ist das Interpolationspolynom über Tschebyscheff-Stütz-
stellen, bei der Wahl $m = n+[\frac{1}{2}n]+1$ ergibt sich ein von Lewanowicz [7] empfohlener Operator. $S_n^{(3)}[f]$ mit $m=n+1$ ist das "Äqui-
oszillationspolynom" zu den Stützstellen $\xi_\kappa = \cos \kappa\pi m^{-1}$
($\kappa=0,1,\ldots,m$), d.h. dasjenige $p \in P_n$, das Proximum für f im Sinne
der sup-Norm ist, sofern der Definitionsbereich auf
$\{\xi_0,\xi_1,\ldots,\xi_m\}$ restringiert wird. Die Wahl $m=2n+1$ gibt ebenfalls einen interessanten Operator $S^{(3)}$ (Lewanowicz [7]).

Für die meisten dieser Operatoren liegen umfangreiche numerische Erfahrungen vor, die zeigen, daß diese Operatoren in der
praktischen Approximationstheorie vorzüglich brauchbar sind.
Eine gewisse theoretische Bestätigung dieser Erfahrungen gibt

Satz 2   Ist S_n eine Folge von Standard-Verfahren, so gilt für
jedes feste k

$$\| f - S_n[f] \| \;\leq\; \rho_{n,n-k} \, \| f^{(n-k)} \| \, (1+o(1)).$$

Der o - Term gilt gleichmäßig in m.

Es möge noch hinzugefügt werden, daß

$$\rho_{n,n+1} = \frac{1}{2^n(n+1)!}$$

lange bekannt und durch Interpolation über Tschebyscheff-Knoten leicht zu beweisen ist. Satz 2 gilt ohne den o - Term auch in dieser Situation (Braß [2], Phillips/Taylor [8], Braß [3]).

2.   Hilfssätze

__Lemma 1.__   Sei k fest. Für $n \to \infty$ gilt

$$\int_{-1}^{1} | D^k[(1-x^2)^{n-1/2}] | \, dx = (\sqrt{n})^{k-1} c_k \, (1+o(1)).$$

Beweis: Durch die Substitution $x\sqrt{n} = y$ erhält man

$$\int_{-1}^{1} | D^k[(1-x^2)^{n-1/2}] | \, dx = (\sqrt{n})^{k-1} \int_{-\sqrt{n}}^{\sqrt{n}} | D^k[(1- \tfrac{y^2}{n})^{n-1/2}] | \, dy \; .$$

Zum Beweis ist also nur zu zeigen, daß die Folge

$$f_n(y) = \begin{cases} | D^k[(1- \tfrac{y^2}{n})^{n-1/2}] | & |y| < \sqrt{n} \\[2mm] 0 & |y| \geq \sqrt{n} \end{cases}$$

punktweise gegen $e^{-y^2} |H_k(y)|$ konvergiert, und daß eine integrierbare Funktion g mit $|f_n(y)| \leq g(y)$ existiert.

Mittels Induktion nach k beweist man, daß

$$D^k[(1-\frac{y^2}{n})^{n-1/2}] = (1-\frac{y^2}{n})^{n-\frac{1}{2}-k} \, P_{k,n}(y)$$

mit durch die Rekursion

$$P_{k+1,n}(y) = -\frac{2n-1-2k}{n} \, y \, P_{k,n}(y) + (1-\frac{y^2}{n}) \, P'_{k,n}(y)$$

$$P_{o,n}(y) = 1$$

bestimmten Polynomen gilt. Aus dieser Formel liest man ab, daß $P_{k,n}$ ein Polynom vom Grad k ist, dessen Koeffizienten rationale Funktionen von n sind, wobei der Zählergrad den Nennergrad nicht übersteigt. Somit existieren die Grenzwerte

$$P_k(y) = \lim_{n\to\infty} P_{k,n}(y)$$

und genügen der Rekursion

$$P_{k+1}(y) = -2y \, P_k(y) + P'_k(y) \, .$$

Vergleich mit einer Formel für Hermite-Polynome (Szegö [11] S. 102, 5.5.10) gibt

$$P_k(y) = (-1)^k \, H_k(y) \, .$$

Wegen

$$(1-\frac{y^2}{n})^{n-\frac{1}{2}-k} \;\to\; e^{-y^2}$$

ist damit die Aussage über die Konvergenz von $f_n(y)$ bewiesen. Da die $P_{k,n}$ nach dem oben Bemerkten Koeffizienten haben, die gleichmäßig in n beschränkt sind, gibt es Polynome $\tilde{P}_k \in P_k$ mit $|P_{k,n}(y)| \leq \tilde{P}_k(y) \, .$

Beachtet man nun noch

$$(1 - \frac{y^2}{n})^n \leq e^{-y^2},$$

also

$$(1 - \frac{y^2}{n})^{n-\frac{1}{2}-k} < e^{-y^2/2}$$

für genügend große n, dann erkennt man, daß

$$g(y) = e^{-y^2/2} \; \tilde{P}_k(y)$$

als Majorante möglich ist.

<u>Lemma 2.</u> Sei k fest. Für $n\to\infty$ gilt

$$\sup_{\|f^{(n-k)}\| < 1} \left| \int_{-1}^{1} f(x) \frac{T_n(x)}{\sqrt{1-x^2}} \, dx \right| = c_k \frac{\sqrt{\pi} \; \sqrt{n}^k}{2^n \, n!} (1+o(1)).$$

Beweis: Mit Hilfe der Rodriguez-Formel für die Tschebyscheff-Polynome (Tricomi [12] S. 188) erhält man

$$\int_{-1}^{1} f(x) \frac{T_n(x)}{\sqrt{1-x^2}} \, dx = \frac{(-1)^n}{(2n-1)!!} \int_{-1}^{1} f(x) \, D^n[(1-x^2)^{n-1/2}] \, dx \; .$$

Durch (n-k)-malige partielle Integration wird das in

$$\frac{(-1)^k}{(2n-1)!!} \int_{-1}^{1} f^{(n-k)}(x) \, D^k[(1-x^2)^{n-1/2}] \, dx$$

umgeformt, und somit hat man

$$\sup_{\|f^{(n-k)}\| \leq 1} \left| \int_{-1}^{1} f(x) \frac{T_n(x)}{\sqrt{1-x^2}} \, dx \right| = \frac{1}{(2n-1)!!} \int_{-1}^{1} |D^k[(1-x^2)^{n-1/2}]| \, dx \; .$$

Die Behauptung folgt nun aus Lemma 1 unter Verwendung der Wallisschen Formel.

<u>Lemma 3.</u> Es gilt

$$E_n[f] \geq \left| \frac{2}{\pi} \int_{-1}^{1} f(x) \frac{T_{n+1}(x)}{\sqrt{1-x^2}}\, dx \right| - \left(1+\frac{4}{\pi}\right) E_{3n+2}[f]$$

Beweis: Das Funktional

$$B[f] = \frac{1}{n+1} \sum_{\nu=0}^{n+1}{}'' (-1)^{\nu}\, f\left(\cos \frac{\nu\pi}{n+1}\right)$$

genügt der Beziehung

$$(3) \qquad\qquad E_n[f] \geq |B[f]|$$

und hat die Eigenschaft

$$B[T_\nu] = 0 \qquad\qquad \nu = 0,1,\ldots,n,\ n+2,\ldots,\ 3n+2$$
$$B[T_{n+1}] = 1$$

(siehe Rivlin [9] S. 142). Also ist das Funktional

$$B - A_1 [\cdot T_{n+1}]$$

Null auf allen $p \in P_{3n+2}$. Ist $p \in P_{3n+2}$ Proximum an f, so folgt

$$|B[f] - A_1[fT_{n+1}]| = |B[f-p] - A_1[(f-p)T_{n+1}]|$$

$$\leq \left(1+\frac{4}{\pi}\right) E_{3n+2}[f] \ .$$

Aus dieser Beziehung folgt wegen (3) die Behauptung.

<u>Lemma 4.</u> $\displaystyle \sup_{\|f^{(n-k)}\| \leq 1} \| f - S_n[f] \| \leq$

$$\frac{1}{2(n-k)} \sup_{\|g^{(n-k-1)}\| \leq 1} \left\{ |A_i[gT_n]| + |A_i[g\,T_{n+1}]| \right\}.$$

Beweis:

$$S_n[f](x) = \sum_{\nu=0}^{n}{}' A_i[fT_\nu]T_\nu(x) = A_i[f \sum_{\nu=0}^{n}{}' T_\nu T_\nu(x)] =$$

$$\frac{1}{2} A_i [f \; \frac{T_{n+1}(x)T_n - T_n(x)T_{n+1}}{x - \cdot}] \; ,$$

wobei von der Christoffel-Darboux-Formel Gebrauch gemacht wurde
(Rivlin [9] S. 35). Wegen

$$f(x)-S_n[f](x) = S_n[f(x)-f](x)$$

folgt

$$f(x)-S_n[f](x) = \frac{1}{2} A_i[\frac{f(x)-f}{x-\cdot} (T_{n+1}(x) \; T_n-T_n(x)T_{n+1})].$$

Definiert man eine Funktion g durch

$$g(t) = \frac{f(x)-f(t)}{x-t} \; ,$$

so hat man also

$$|f(x)-S_n[f](x)| = |\frac{1}{2} T_{n+1}(x) \; A_i[gT_n] - \frac{1}{2} T_n(x) \; A_i[gT_{n+1}]|$$

$$\leq \frac{1}{2} \{|A_i[gT_n]| + |A_i[gT_{n+1}]|\} \; .$$

Beachtet man noch die Beziehung

$$\|g^{(n-k-1)}\| \leq \frac{\|f^{(n-k)}\|}{n-k}$$

(siehe Braß/Schmeißer [5]), so folgt die Behauptung.

3. Die Beweise

Beweis von Satz 1. S_n bedeute hier den auf A_1 gegründeten Standardoperator. Kombiniert man Lemma 2 und Lemma 4, so folgt sofort

$$(4) \qquad \sup_{\|f^{(n-k)}\| \leq 1} \|f - S_n[f]\| \leq \frac{c_{k+1}}{\sqrt{\pi}} \, \frac{(\sqrt{n})^{k+1}}{2^n (n+1)!} \, (1+o(1)).$$

Hiermit hat man eine Abschätzung von $E_n[f]$ nach oben. Zur Abschätzung in der umgekehrten Richtung zieht man Lemma 3 heran. Nach Sinwels [10] Form des Jacksonschen Satzes ist

$$E_n[f] \leq \frac{\pi}{2} \, \frac{(n+1-r)!}{(n+1)!} \, \|f^{(r)}\|$$

also

$$E_{3n+2}[f] \leq \frac{\pi}{2} \, \frac{(2n+3+k)!}{(3n+3)!} \, \|f^{(n-k)}\| .$$

Damit gibt Lemma 3 unter Beachtung von Lemma 2

$$(5) \qquad \sup_{\|f^{n-k)}\| \leq 1} E_n[f] \geq$$

$$\frac{c_{k+1}}{\sqrt{\pi}} \, \frac{(\sqrt{n})^{k+1}}{2^n (n+1)!} \, (1+o(1)) - 0 \left(\frac{(2n+3+k)!}{(3n+3)!} \right) .$$

Anwendung der Stirlingschen Formel zeigt, daß der 0-Term in den o-Term des ersten Summanden hineingezogen werden kann.

(4) und (5) ergeben nun die Behauptung.

Beweis von Satz 2.   Derjenige Spezialfall, in dem $S_n[f]$ die Teilsumme der Tschebyscheff-Entwicklung bedeutet, ist durch den vorangehenden Beweis erledigt. Sei nun S_n der auf A_i $(i=2,3)$ basierende Operator. Ausgangspunkt ist die Quadraturformel

$$A_1[p] = A_i[p] , \qquad i=2,3$$

gültig für alle $p \in P_{2m-1}$ (Rivlin [9] S. 41, S. 50). Ist $p \in P_{2m-1}$ das Proximum für f, so folgt

$$|A_1[f] - A_i[f]| = |A_1[f-p] - A_i[f-p]| \leq (\tfrac{4}{\pi} + 2)\|f-p\|$$

$$\leq 4\, E_{2m-1}[f].$$

Weiter ist dann

$$|A_1[fT_n] - A_i[fT_n]| \leq 4E_{2m-1}[fT_n] \leq 4E_{2m-n-1}[f] \leq 4E_n[f].$$

Zieht man nun die Folgerung aus Satz 1 sowie Lemma 2 heran, so erhält man

$$\sup_{\|f^{(n-k-1)}\| \leq 1} |A_i[fT_n]| \leq \frac{c_{k+1}}{\sqrt{\pi}}\ \frac{(\sqrt{\pi})^{k+1}}{2^{n-1}\, n!}\ (1+o(1))$$

und ganz analog

$$\sup_{\|f^{(n-k-1)}\| \leq 1} |A_i[fT_{n+1}]| \leq \mathrm{const}_k\ \frac{(\sqrt{n})^k}{2^n\, n!} .$$

Setzt man diese Beziehungen in Lemma 4 ein, so folgt Satz 2.

4. Eine Anwendung

Exemplarisch für die möglichen Anwendungen sei hier ein Problem der Restabschätzung zum Clenshaw-Curtis-Quadraturverfahren (siehe z.B. Braß [4])betrachtet. Der Rest der Clenshaw-Curtis-Formel mit n Stützstellen werde mit R_n bezeichnet; sofern n ungerade ist, existiert eine Zahl d_n mit

$$|R_n[f]| \leq d_n \, \|f^{(n+1)}\| \; .$$

Die genaue Bestimmung der kleinsten zulässigen Zahlen d_n ist hier schwieriger als bei den Quadraturformeln von Gauß, Newton-Cotes und Romberg, die alle definit sind; die Clenshaw-Curtis-Formeln sind für $n > 3$ nicht definit (Akrivis/Förster [1]).

$S_n[f]$ bezeichne die n-te Teilsumme der Tschebyscheff-Entwicklung von f. Dann gilt

$$|R_n[f]| \;\leq\; |R_n[f-S_{n+7}[f]] \;+\; \sum_{\nu=n+1}^{n+7} a_\nu \, R_n[T_\nu]|$$

$$\leq \|R_n\| \; \|f-S_{n+7}[f]\| + |R_n[T_{n+1}]| \; \{|a_{n+1}| + $$

$$\sum_{\nu=n+2}^{n+7} |a_\nu| \; \frac{R_n[T_\nu]}{R_n[T_{n+1}]} \} \; .$$

Beachtet man nun (Braß [4])

$$\|R_n\| = 4 \; , \; R_n[T_{n+1}] \;=\; \frac{16(n-1)}{(n-4)(n^2-4)n} \; ,$$

$$R_n[T_\nu] < \frac{const}{n^3} \qquad \nu=n+2, \, n+3, \ldots, \, n+7,$$

so hat man

$$| R_n[f] | \leq 4 \| f - S_{n+7}[f] \| + | R_n[T_{n+1}] | \{ |a_{n+1}| + \text{const} \sum_{\nu=n+2}^{n+7} |a_\nu| \}.$$

Weiter ist

$$|a_\nu| \leq \text{const} \; E_{\nu-1}[f]$$

und (Braß [2])

$$a_{n+1} = 2^{-n}(n+1)!^{-1} \, f^{(n+1)}(\xi).$$

Setzt man das ein und beachtet Satz 1 und Satz 2, so erkennt man, daß der Term $|a_{n+1} \, R_n[T_{n+1}]|$ die übrigen majorisiert; damit ist bewiesen

$$(6) \qquad d_n = \frac{R_n[T_{n+1}]}{2^n \, (n+1)!} \; (1+o(1)) \; ,$$

und am Beispiel $f = T_{n+1}$ erkennt man sofort, daß dieser d_n-Wert unverbesserbar ist.

Für einzelne n-Werte lassen sich die optimalen d_n mittels Peanokerntheorie bestimmen; die folgende Tabelle von asymptotischen ((6) ohne o-Term) und (von Herrn cand. math. M. Nasgowitz bestimmten) optimalen d_n-Werten zeigt eine gute Übereinstimmung.

$n =$	5	7	9
d_n^{as}	2,64550	1,96838	1,98826
d_n^{opt}	2,66769	1,96863	1,98826
	$\cdot \; 10^{-5}$	$\cdot \; 10^{-8}$	$\cdot \; 10^{-11}$

Literatur

[1] Akrivis, G., Förster, K.-J.: Persönliche Mitteilung

[2] Braß, H.: Approximation durch Teilsummen von Orthogonal-
polynomreihen. In: Numerical Methods of Approximation
Theory, Edited by L. Collatz, G. Meinardus, H. Werner;
Birkhäuser 1980, S. 69 - 83.

[3] Braß, H.: Error Estimates for Least Squares Approximation
by Polynomials. Erscheint in J. Approximation Theory.

[4] Braß, H.: Quadraturverfahren. Göttingen-Zürich, Vandenhoeck
und Ruprecht 1977.

[5] Braß, H., Schmeisser, G.: Error Estimates for Interpolatory
Quadrature Formulae. Numer. Math. 37, S. 371-386 (1981).

[6] Ehresmann, D.: Zur Abschätzung des Bestapproximationsfeh-
lers bei der Approximation differenzierbarer Funktionen
durch Polynome. Numer. Math. 13, S. 94 - 100 (1969).

[7] Lewanowicz, S.: Some Polynomial Projections with Finite
Carrier. J. Approximation Theory 34, S. 249 - 263 (1982).

[8] Phillips, G.M., Taylor, P.J.: Polynomial Approximation
Using Equioscillation on the Extreme Points of Chebyshev
Polynomials. J. Approximation Theory 36, S. 257 - 264
(1982).

[9] Rivlin, Th.J.: The Chebyshev Polynomials. New York, Wiley
1974.

[10] Sinwel, H.F.: Uniform Approximation of Differentiable
Functions by Algebraic Polynomials. J. Approximation Theory
32, S. 1 - 8 (1981).

[11] Szegö, G.: Orthogonal Polynomials. New York, Amer. Math.
Soc. 1939.

[12] Tricomi, F.G.: Vorlesungen über Orthogonalreihen. Berlin,
Springer 1970.

Prof. Dr. Helmut Braß
Institut für Angewandte Mathematik
der Technischen Universität Braunschweig,
3300 Braunschweig, Pockelsstraße 14, BRD

ISNM, Vol. 67
Numerical Methods of
Approximation Theory, Vol. 7
© 1983 Birkhäuser Verlag Basel

ALGORITHMS TO COMPUTE THE REFLECTION COEFFICIENTS OF DIGITAL FILTERS

Adhemar Bultheel

Three algorithms are given which compute explicitly or implicitly certain coefficients in a recursive scheme to generate polynomials. It is shown how these algorithms solve some Padé-like approximation problems. As a special case of these algorithms we obtain the Levinson and Schur methods to compute the reflection coefficients in linear predictive filtering. The third algorithm generates a table of numbers related to the reflection coefficients by certain rhombus rules. By some numerical examples we show how we can obtain information about the stability properties (pole location) and zero structure (transmission zeros) of the transfer-function of the system from all of these three algorithms.

1. Situation of the Problem and a practical example

In [10] the authors gave a relation between continued fractions and digital filters. The algorithms were mainly applied to obtain information of the stability properties of the digital filter. I.e. on the location of its poles. We shall continue in this paper to investigate the relation that exists between digital filters, continued fractions, Laurent-Padé approximations etc. New is the information that we shall obtain about the location of the transmission zeros. We shall start by introducing a practical example which motivates the formulation of the mathematical problem given lateron

We shall consider the problem of giving a model for human speech production (see [11,12]. In fig. 1 we give a discretized model of the vocal tract. It is a sequence of coaxial cylinders where at the left we think the glottis and at the right the lips. A sound such as a vowel is produced by a pule train produced by the vocal chords (input). This is modeled by the vocal tract and a periodic soundwave is produced at the lips (output). If we sample this output we can think of this model as a discrete system (fig. 2). There is some discretized input u_n. This is convolved with an impulse response h_k and out comes a discretized y_n. Instead of working with sequences we prefer to work with their z-transforms. e.g. $U(z) = \Sigma\, u_k\, z^k$. Convolution goes over into multiplication and thus $Y(z) = H(z)\, U(z)$. $H(z)$ is called the <u>transfer function</u> of the system. Suppose we want to find an approximation for $H(z)$ which is of the form $c/\sum_{k=0}^{n} q_k\, z^k$ (with $q_0=1$). This is called an <u>AR</u> (Autoregressive) <u>filter</u>. That is

$$y_t = c\, u_t - y_{t-1}\, q_1 - \cdots - y_{t-n}\, q_n \tag{1}$$

Thus we predict the output at time t from the n previous outputs and the input at time t. Hence this is called a <u>predictive filter</u>. There are some physical restrictions on $H(z)$. First it has to be <u>causal</u>. That is there can be no output before there has been an input. If $U(z) = 1$ (i.e. $u_t = \delta_{t0}$) then $Y(z) = H(z)$ and thus h_t for $t < 0$ must be zero. Thus $H(z)$ can not contain negative powers of z. Secondly $H(z)$ must be <u>stable</u>. This means that the energy of the output must be bounded if the energy of the input is bounded. Thus the power spectrum of H must be bounded i.e. $\sum_{0}^{\infty} |h_k|^2 < \infty$. Consequently $H(z)$ can have no poles inside the unit disc of the complex plane. A least squares solution of the infinite dimensional system (1) is given by the solution of its normal equations. Suppose the input is $u_t = \delta_{t0}$, and thus $y_t = h_t$ then the normal equations are

$$\sum_{j=1}^{n} f_{k-j}\, q_j = \delta_{k0}\, c^2 \qquad k = 0,1,\ldots,n \qquad (q_0=1)$$

where $f_k = \sum_{j=0}^{\infty} h_{k-j}\, h_j \quad k = 0,\pm1,\pm2,\ldots$ are the <u>autocorrelation coefficients.</u> It is clear that $F(e^{i\theta}) = \sum_{-\infty}^{\infty} f_k\, e^{ik\theta} = |H(e^{i\theta})|^2$. We shall refer to $F(e^{i\theta})$ as the <u>power spectrum</u>. Thus to obtain a least squares predictive filter we have to solve the following mathematical problem : Given a function F such that

$$F(e^{i\theta}) = F(e^{-i\theta}) \qquad \text{(even)} \tag{F1}$$
$$F(e^{i\theta}) \geqslant 0 \qquad \text{(nonnegative)} \tag{F2}$$

find $H(e^{i\theta})$ such that $\quad F(e^{i\theta}) = |H(e^{i\theta})|^2 \tag{H1}$

$$H(z) \in H_2 \quad \text{(Hardy Space of the unit disc).} \tag{H2}$$

This formulation is also applicable for general ARMA (Autoregressive-moving average) filters. I.e. where the approximation for H is a general rational function. Since in F we have only information on the square of H we should decide in the general ARMA case where we want the zeros of H to make the problem uniquely solvable. Therefore we add a third condition on H viz.

H should be outer of H_2 [15]   (H3)

We have the following existence result.

<u>THEOREM 1</u> [6]

Let F satisfy the conditions F1, F2 and $\log F \in L_1$ (i.e. Lebesgue integrable on the unit circle) then there exists a function H such that $F(e^{i\theta}) = |H(e^{i\theta})|^2$ which is given by

$$H(e^{i\theta}) = \lim_{r \nearrow 1} \exp \{\frac{1}{4\pi} \int_{-\pi}^{\pi} \frac{e^{i\varphi}+re^{i\theta}}{e^{i\varphi}-re^{i\theta}} \log F(e^{i\theta})d\theta\}$$

H is in H_2 and is outer. Its Taylor series

$$H(z) = \sum_0^\infty h_k z^k \quad \text{converges in} \quad |z| < 1 \quad \text{and} \quad h_0 > 0.$$

Under conditions of roomyness [6], H(z) can be defined in all of $\mathbb{C}$ and $H(z).\overline{H(1/\bar{z})}$ is an extension of F in $\mathbb{C}$. We denote this also by F(z). We shall now have to construct an approximation for H which is rational of degree n, which also satisfies H2 and H3 and which satisfies H1 as good as possible. As n tends to infinity this should converge to H(z).
The further outline of the paper is as follows : In section 2 we introduce algorithms A1, A2 and A3 and in section 3 we shall show how these solve some approximation problems in the Padé sphere. Section 4 shows how the above AR-filtering problem appears as a special case and how the approximants have some of the desired properties. Section 5 gives the relevance of a qd-like algorithm for the computation of zeros of H(z), i.e. the transmission zeros. Section 6 gives some numerical example to illustrate section 5.

2. Algorithms

ALGORITHM A1

First we introduce a fast algorithm to solve Toeplitz systems of linear equations.
Let $\{f_k\}_{-\infty}^\infty$ be a normal sequence, i.e. all the Toeplitz matrices

$T_n^{(m)} = [f_{m+k-j}]_{k,j=0}^n$ are nonsingular. Suppose you want to solve the couple of linear equations $T_n^{(m)} Q_n^{(m)} = [u_n^{(m)} \ 0 \ldots 0]^T$ and $T_n^{(m)} \hat{Q}_n^{(m)} = [0 \ldots 0 \ \hat{v}_n^{(m)}]^T$ for $n = 0,1,2,\ldots$

Let $Q_n^{(m)} = [q_{0,n}^{(m)} = 1, q_{1,n}^{(m)}, \ldots, q_{n,n}^{(m)}]^T$ and $\hat{Q}_n^{(m)} = [\hat{q}_{0,n}^{(m)}, \ldots, \hat{q}_{n,n}^{(m)} = 1]^T$ and define the polynomials

$$Q_n^{(m)}(z) = [1 \ z \ldots z^n] Q_n^{(m)} \quad \text{and} \quad \hat{Q}_n^{(m)}(z) = [1 \ z \ldots z^n] \hat{Q}_n^{(m)}.$$

Then it is well known that these satisfy the recursion [14]

$$[Q_{n+1}^{(m)}(z) \ \hat{Q}_{n+1}^{(m)}(z)] = [Q_n^{(m)}(z) \ \hat{Q}_n^{(m)}(z)] \begin{bmatrix} 1 & -\beta_{n+1}^{(m)} \\ & \\ -z\alpha_{n+1}^{(m)} & z \end{bmatrix}$$

with $\alpha_{n+1}^{(m)} = v_n^{(m)} / \hat{v}_n^{(m)}$, $\beta_{n+1}^{(m)} = \hat{u}_n^{(m)} / u_n^{(m)}$, $v_n^{(m)} = [f_{m+n+1} \cdots f_{m+1}] Q_n^{(m)}$,

$\hat{u}_n^{(m)} = [f_{m-1} \cdots f_{m-n-1}] \hat{Q}_n^{(m)}$ and $u_{n+1}^{(m)} = \hat{v}_{n+1}^{(m)} = u_n^{(m)} (1 - \alpha_{n+1}^{(m)} \beta_{n+1}^{(m)})$, with

initial conditions $Q_0^{(m)} = \hat{Q}_0^{(m)} = 1$, $u_0^{(m)} = \hat{v}_0^{(m)} = f_m$.

We shall look at this algorithm as a method to generate the numbers $\alpha_n^{(m)}$ and $\beta_n^{(m)}$ for constant m from the data $\{f_k\}$. Remark that the constant term of $\hat{Q}_n^{(m)}(z)$ is $-\beta_n^{(m)}$ and that $Q_n^{(m-1)}(z) = -\hat{Q}_n^{(m)}(z)/\beta_n^{(m)}$. The coefficient of z^n in $Q_n^{(m)}(z)$ is $-\alpha_n^{(m)}$.

ALGORITHM A2

Define a projection operation for formal Laurent series :

$$\Pi_{a:b}\left(\sum_{-\infty}^{\infty} s_k z^k\right) = \sum_a^b s_k z^k.$$

Furthermore we introduce the notations O_+ and O_- as follows. Any Laurent-series $S(z)$ with $\Pi_{-\infty:m+1} S(z) = 0$ is denoted by $O_+(z^m)$ and if $\Pi_{m+1:\infty} S(z) = 0$, then by $O_-(z^m)$. With the same notation as in section 2.1 we can introduce formally : $F(z) = \sum_{-\infty}^{\infty} f_k z^k$,

$$P_n^{(m)}(z) = \Pi_{-\infty:m} (F(z) Q_n^{(m)}(z)) = u_n^{(m)} z^m + O_-(z^{m-1})$$

$$R_n^{(m)}(z) = \Pi_{m+1:\infty} (F(z) Q_n^{(m)}(z)) = v_n^{(m)} z^{m+n+1} + O_+(z^{m+n+2})$$

$$\hat{P}_n^{(m)}(z) = \Pi_{-\infty:m-1}(F(z)\,\hat{Q}_n^{(m)}(z)) = \hat{u}_n^{(m)}\,z^{m-1} + O_-(z^{m-2})$$

$$\hat{R}_n^{(m)}(z) = \Pi_{m:\infty}(F(z)\,\hat{Q}_n^{(m)}(z)) = \hat{v}_n^{(m)}\,z^{m+n} + O_+(z^{m+n+1})$$

It is not difficult to see that $(P,\hat{P})$ and $(R,\hat{R})$ obey the same recursions as $(Q,\hat{Q})$. Thus

$$[\,S_{n+1}^{(m)}(z)\ \ \hat{S}_{n+1}^{(m)}(z)\,] = [\,S_n^{(m)}(z)\ \ \hat{S}_n^{(m)}(z)\,]\begin{bmatrix} 1 & -\beta_{n+1}^{(m)} \\ & \\ -z\alpha_{n+1}^{(m)} & z \end{bmatrix} \tag{R}$$

where S can be P, Q or R.

The initial conditions are $P_0^{(m)} = \sum\limits_{-\infty}^{m} f_k\,z^k$, $R_0^{(m)} = \sum\limits_{m+1}^{\infty} f_k\,z^k$,

$\hat{P}_0^{(m)} = \sum\limits_{-\infty}^{m-1} f_k\,z^k$ and $\hat{R}_0^{(m)} = \sum\limits_{m}^{\infty} f_k\,z^k$. Since the leading terms in P, R, $\hat{P}$ and $\hat{R}$ give us the numbers u, $\hat{u}$, v and $\hat{v}$, we can generate $\alpha_n^{(m)}$ and $\beta_n^{(m)}$ for constant m by executing the above recursion for S = P and S = R in parallel. This is thus an alternative for algorithm A1. This algorithm A2 is also found in [14].

ALGORITHM A3

Define the numbers $b_k^{(m)} = -\alpha_{k-1}^{(m)}\,\beta_k^{(m)}$ and $a_k^{(m)} = (1-\alpha_{k-1}^{(m)}\,\beta_{k-1}^{(m)})\alpha_{k-2}^{(m)}/\alpha_{k-1}^{(m)}$ for $k \geqslant 2$, $m = 0,\pm1,\pm2,\ldots$

It is possible to prove that these numbers satisfy the rhombus rules [2]

$$a_k^{(m)}\,b_{k-1}^{(m+1)} = a_k^{(m+1)}\,b_k^{(m+1)} \qquad\qquad k \geqslant 2$$

$$a_{k+1}^{(m)} + b_k^{(m+1)} = b_k^{(m)} + a_k^{(m)} \qquad\qquad k \geqslant 1$$

with initial conditions

$$b_k^{(m+1)} = f_m/f_{m+1}$$
$$a_1^{(m+1)} = 0 \qquad\qquad m = 0,\pm1,\pm2,\ldots$$

The proof is like the proof for the Rutishauser qd rules. It can be given via continued fractions (see next section) or via the expressions of u, v, $\hat{u}$

and $\hat{v}$ in terms of Toeplitz determinants. This algorithm is closely related to the qd algorithm [9], the $\pi\zeta$ algorithm [7] and the FG algorithm [11]. We can by this algorithm construct a table

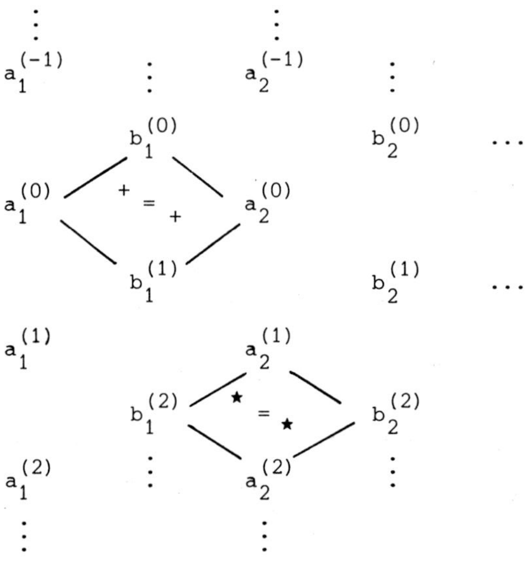

Once this table is known, we can recover the α and β numbers from it easily so that it is another alternative for algorithm A1 and algorithm A2.

3. Approximations, Moebius transforms and continued fractions

3.1. Padé approximants

From their definitions we find that

$$F(z)\, Q_n^{(m)}(z) - P_n^{(m)}(z) = R_n^{(m)}(z) = O_+(z^{m+n+1})$$

and

$$F(z)\, \hat{Q}_n^{(m)}(z) - \hat{P}_n^{(m)}(z) = \hat{R}_n^{(m)}(z) = O_+(z^{m+n}).$$

Since $Q_n^{(m)}(z)$ and $\hat{Q}_n^{(m)}(z)$ are polynomials of degree n and because $P_n^{(m)}(z) = O_-(z^n)$ and $\hat{P}_n^{(m)}(z) = O_-(z^{n-1})$, we have here a kind of a generalized Padé approximants. In the first case of type (m/n) and in the second case of type $(m-1/n)$.

3.2. Laurent-Padé approximants

Define $Z^{(m)}(z) = \frac{1}{2} f_m z^m + f_{m+1} z^{m+1} + \ldots$ and

$$\hat{Z}^{(m)}(z) = -\frac{1}{2} f_m z^m - f_{m-1} z^{m-1} - \ldots$$

Then $F(z) = Z^{(m)}(z) - \hat{Z}^{(m)}(z)$.

Because $F(z) Q_n^{(m)}(z) - P_n^{(m)}(z) = R_n^{(m)}(z)$

we have

$$Z^{(m)}(z) Q_n^{(m)}(z) - R_n^{(m)}(z) = \hat{Z}^{(m)}(z) Q_n^{(m)}(z) + P_n^{(m)}(z).$$

The left hand side is $O_+(z^m)$ and the right hand side is $O_-(z^{m+n})$. Thus both are of the form $z^m A_n^{(m)}(z)$ with $A_n^{(m)}(z)$ a polynomial of degree n. Similarly define the polynomial : $\hat{A}_n^{(m)}(z)$ of degree n by

$$z^m \hat{A}_n^{(m)}(z) = Z^{(m)}(z) \hat{Q}_n^{(m)}(z) - \hat{R}_n^{(m)}(z) = \hat{Z}_n^{(m)}(z) \hat{Q}_n^{(m)}(z) + \hat{P}_n^{(m)}(z).$$

Since

$$\begin{bmatrix} P_{n+1}^{(m)}(z) & \hat{P}_{n+1}^{(m)}(z) \\ Q_{n+1}^{(m)}(z) & \hat{Q}_{n+1}^{(m)}(z) \end{bmatrix} = \begin{bmatrix} P_n^{(m)}(z) & \hat{P}_n^{(m)}(z) \\ Q_n^{(m)}(z) & \hat{Q}_n^{(m)}(z) \end{bmatrix} \begin{bmatrix} 1 & -\beta_{n+1}^{(m)} \\ -z\alpha_{n+1}^{(m)} & z \end{bmatrix}$$

by multiplying from the left with $[1 \; Z^{(m)}(z)]$ we obtain again the recursion (R) for the pair $(A,\hat{A})$. The initial conditions are : $A_0^{(m)}(z) = f_m/2$ and $\hat{A}_0^{(m)} = -f_m/2$. We thus have

$$\begin{bmatrix} \frac{1}{2} f_m & -\frac{1}{2} f_m \\ 1 & 1 \end{bmatrix} \begin{bmatrix} 1 & -\beta_1^{(m)} \\ -z\alpha_1^{(m)} & z \end{bmatrix} \cdots \begin{bmatrix} 1 & -\beta_n^{(m)} \\ -z\alpha_n^{(m)} & z \end{bmatrix} = \begin{bmatrix} A_n^{(m)}(z) & \hat{A}_n^{(m)}(z) \\ Q_n^{(m)}(z) & \hat{Q}_n^{(m)}(z) \end{bmatrix}.$$

Take the determinant and you find that

$$A_n^{(m)}(z) \hat{Q}_n^{(m)}(z) - \hat{A}_n^{(m)}(z) Q_n^{(m)}(z) = z^n f_m \prod_1^n (1-\alpha_k^{(m)} \beta_k^{(m)}).$$

Thus

$$\frac{A_n^{(m)}(z)}{Q_n^{(m)}(z)} - \frac{\hat{A}_n^{(m)}(z)/z^n}{\hat{Q}_n^{(m)}(z)/z^n} = \frac{f_m \prod_1^n (1-\alpha_k^{(m)} \beta_k^{(m)})}{(Q_n^{(m)}(z))(\hat{Q}_n^{(m)}(z)/z^n)}$$

By definition is this a (0/n) Laurent-Padé approximant of $F(z)/z^m$ [8].

3.3. Two-point Padé approximants

From the definition of $A_n^{(m)}(z)$ and $\hat{A}_n^{(m)}(z)$ we get

$$z^{-m}\,\hat{Z}^{(m)}(z)\,Q_n^{(m)}(z) - A_n^{(m)}(z) = -z^{-m}\,P_n^{(m)}(z) = 0_-(z^0)$$

$$z^{-m}\,Z^{(m)}(z)\,\hat{Q}_n^{(m)}(z) - \hat{A}_n^{(m)}(z) = -z^{-m}\,R_n^{(m)}(z) = 0_+(z^n)$$

$$z^{-m}\,Z^{(m)}(z)\,Q_n^{(m)}(z) - A_n^{(m)}(z) = z^{-m}\,R_n^{(m)}(z) = 0_+(z^{m+1})$$

$$z^{-m}\,\hat{Z}^{(m)}(z)\,\hat{Q}_n^{(m)}(z) - \hat{A}_n^{(m)}(z) = z^{-m}\,\hat{P}_n^{(m)}(z) = 0_-(z^{-1}).$$

Denote by $L_+\,(A_n^{(m)}(z)/Q_n^{(m)}(z))$ the formal expansion in positive powers of z for the ratio of the two polynomials and similarly by L_- the negative power expansion. Then clearly

$$z^{-m}\,Z^{(m)}(z) - L_+(A_n^{(m)}(z)/Q_n^{(m)}(z)) = z^{-m}\,R_n^{(m)}(z)/Q_n^{(m)}(z) = 0_+(z^{n+1})$$

$$z^{-m}\,\hat{Z}^{(m)}(z) - L_-(\hat{A}_n^{(m)}(z)/\hat{Q}_n^{(m)}(z)) = -z^{-m}\,\hat{P}_n^{(m)}(z)/\hat{Q}_n^{(m)}(z) = 0_-(z^{-n-1})$$

$$z^{-m}\,Z^{(m)}(z) - L_+(\hat{A}_n^{(m)}(z)/\hat{Q}_n^{(m)}(z)) = z^{-m}\,\hat{R}_n^{(m)}(z)/\hat{Q}_n^{(m)}(z) = 0_+(z^n)$$

$$z^{-m}\,\hat{Z}^{(m)}(z) - L_-(A_n^{(m)}(z)/\hat{Q}_n^{(m)}(z)) = z^{-m}\,P_n^{(m)}(z)/Q_n^{(m)}(z) = 0_-(z^{-n}).$$

Therefore $A_n^{(m)}(z)/Q_n^{(m)}(z)$ and $\hat{A}_n^{(m)}(z)/\hat{Q}_n^{(m)}(z)$ are both two-point Padé approximants for the pair $(z^{-m}\,Z^{(m)}(z),\ z^{-m}\,\hat{Z}^{(m)}(z))$ of appropriate types. [16]

3.4. Moebius transforms

The recursion (R) is related to the Moebius transform.

$$t_n(w) = \frac{w - \alpha_n^{(m)}}{1 - \beta_n^{(m)}\,w}$$

in the sense that

$$\frac{S_n^{(m)}(z)}{\hat{S}_n^{(m)}(z)} = t_n\left(\frac{S_{n-1}^{(m)}(z)}{z\,\hat{S}_{n-1}^{(m)}(z)}\right)$$

From this we can derive e.g. the following recursions :

Set $\Gamma_n^{(m)}(z) = \dfrac{R_n^{(m)}(z)}{z \, \hat{R}_n^{(m)}(z)}$, then $\Gamma_n^{(m)}(z) = \dfrac{1}{z} \dfrac{\Gamma_{n-1}^{(m)}(z) - \alpha_n^{(m)}}{1 - \beta_n^{(m)} \Gamma_{n-1}^{(m)}(z)}$, and $\alpha_{n+1}^{(m)}$ is

the leading term in $\Gamma_n^{(m)}(z) = \alpha_{n+1}^{(m)} + 0_+(z)$.

Set $\hat{\Gamma}_n^{(m)}(z) = \dfrac{z \, \hat{P}_n^{(m)}(z)}{P_n^{(m)}(z)}$, then $\hat{\Gamma}_n^{(m)}(z) = z \dfrac{\hat{\Gamma}_{n-1}^{(m)}(z) - \beta_n^{(m)}}{1 - \alpha_n^{(m)} \hat{\Gamma}_{n-1}^{(m)}(z)}$, and $\beta_{n+1}^{(m)}$ is

the leading term in $\hat{\Gamma}_n^{(m)}(z) = \beta_{n+1}^{(m)} + 0_-(z^{-1})$.

Set $\hat{\Pi}_n^{(m)}(z) = L_-\left(\dfrac{Q_n^{(m)}(z)}{\hat{Q}_n^{(m)}(z)}\right)$, then $\hat{\Pi}_{n-1}^{(m)}(z) = z \dfrac{\hat{\Pi}_n^{(m)}(z) + \alpha_n^{(m)}}{1 + \beta_n^{(m)} \hat{\Pi}_n^{(m)}(z)}$, and

$\hat{\Pi}_n^{(m)}(z) = -\alpha_n^{(m)} + 0_-(z^{-1})$.

Set $\Pi_n^{(m)}(z) = L_+\left(\dfrac{\hat{Q}_n^{(m)}(z)}{Q_n^{(m)}(z)}\right)$, then $\Pi_{n-1}^{(m)}(z) = \dfrac{1}{z} \dfrac{\Pi_n^{(m)}(z) + \beta_n^{(m)}}{1 + \alpha_n^{(m)} \Pi_n^{(m)}(z)}$, and

$\Pi_n^{(m)}(z) = -\beta_n^{(m)} + 0_+(z)$.

3.5. <u>Continued fractions</u>

The Moebius transforms of (R) are also related to continued fractions.
Consider the continued fraction

$$c_0 + \cfrac{c_1}{1} + \cfrac{z}{-\beta_1^{(m)}} + \cfrac{1 - \alpha_1^{(m)} \beta_1^{(m)}}{-\alpha_1^{(m)}} + \cfrac{z}{-\beta_2^{(m)}} + \cfrac{1 - \alpha_2^{(m)} \beta_2^{(m)}}{-\alpha_2^{(m)}} + \cdots$$

With initial conditions

a) $(c_0, c_1) = (\hat{P}_0^{(m)}(z), f_m z^m)$ it has convergents $\hat{P}_0^{(m)}(z)/\hat{Q}_0^{(m)}(z)$,
$P_0^{(m)}(z)/Q_0^{(m)}(z)$, $\hat{P}_1^{(m)}(z)/\hat{Q}_1^{(m)}(z),\ldots$ and it can be considered as a formal expansion of $F(z)$.

b) $(c_0, c_1) = (-\frac{1}{2} f_m, f_m)$ it has convergents $\hat{A}_0^{(m)}(z)/\hat{Q}_0^{(m)}(z)$, $A_0^{(m)}(z)/Q_0^{(m)}(z)$,
$\hat{A}_1^{(m)}(z)/\hat{Q}_1^{(m)}(z),\ldots$ and it can be considered as a formal expansion of
$z^{-m} \hat{Z}^{(m)}(z)$ or $-z^{-m} Z^{(m)}(z)$.

Remark that the above continued fraction is equivalent with

$$c_0 \;+\; \frac{c_1}{|\;1\;} \;-\; \frac{z}{|\;b_1^{(m)}} \;-\; \frac{a_2^{(m)}}{|\;1\;} \;-\; \frac{z}{|\;b_2^{(m)}} \;-\; \ldots \;,$$

where the $a_k^{(m)}$ and $b_k^{(m)}$ are as defined in algorithm A3.

Because $\Gamma_0^{(m)}(z) = \dfrac{R_0^{(m)}(z)}{z\,\hat{R}_0^{(m)}(z)} = \dfrac{1}{z}\,\dfrac{z^{-m}\,Z^{(m)} - f_m/2}{z^{-m}\,Z^{(m)} + f_m/2}$ we have also

$$\Gamma_0^{(m)}(z) = \beta_1^{(m)} \;-\; \frac{1 - \alpha_1^{(m)}\,\beta_1^{(m)}}{|\;-\alpha_1^{(m)}} \;+\; \frac{z}{|\;-\beta_1^{(m)}} \;+\; \ldots$$

4. Special case : filtering problem

Because in filtering applications F is even we have $f_k = f_{-k}$ (we suppose f_k real) and this symmetry implies e.g. that

$$Q_n^{(m)}(z) = z^n\,\hat{Q}_n^{(-m)}(1/z)$$

$$\alpha_n^{(m)} = \beta_n^{(-m)}$$

$$R_n^{(m)}(z) = z^n\,\hat{P}_n^{(-m)}(1/z)$$

$$P_n^{(m)}(z) = z^n\,\hat{R}_n^{(-m)}(1/z)$$

$$Z^{(m)}(z) = -\hat{Z}^{(-m)}(1/z)$$

$$A_n^{(m)}(z) = -z^n\,\hat{A}_n^{(-m)}(1/z)$$

$$\Gamma_n^{(m)}(z) = \hat{\Gamma}_n^{(-m)}(1/z) .$$

In case of AR filtering we are interested in the case $m = 0$. We shall drop the superscript m whenever it is zero and introduce a notation for the reciprocal of a polynomial. If $p_n(z)$ is a polynomial of degree n, then we set $p_n^{\star}(z) = z^n\,p_n(1/z)$.

It is well known from the theory of moment problems [1] that

<u>THEOREM 2</u>   The following are equivalent :

(1) $F(e^{i\theta}) \geqslant 0$

(2) $Z(z) = \dfrac{1}{2} f_0 + \sum_1^\infty f_k \, z^k \in C$ (Carathéodory class)

      i.e. $Z(z)$ analytic and $\Re e \, Z(z) \geqslant 0$ in $|z| < 1$

(3) $\Gamma_0(z) = \dfrac{1}{z} \dfrac{Z(z) - f_0/2}{Z(z) + f_0/2} = \dfrac{R_0(z)}{z \, \hat{R}_0(z)} \in S$ (Schur class)

      i.e. $\Gamma_0(z)$ analytic and $|\Gamma_0(z)| \leqslant 1$ in $|z| < 1$

(4) $T_n = [\, f_{k-j}\,]^n_{k,j=0} \geqslant$ for all n $\square$

LEMMA 1

(1)   If $f \in S$ then $t(f) \in S$ if $t(w) = \dfrac{w - \rho}{1 - \bar{\rho} w}$, $|\rho| < 1.$

      If moreover $\rho = f(0)$, then $\dfrac{1}{z} t \, (f(z)) \in S.$ [1 p.102]

(2)   If $f \in S$ then $z \, f(z) \in S$

(3)   If $Z \in C$ then $c(Z) \in S$ if $c(w) = \dfrac{w - \gamma}{w + \bar{\gamma}}$ and $\Re e \, \gamma > 0.$

      If moreover $\gamma = Z(0)$, then also $\dfrac{1}{z} c(Z) \in S.$ [1 p.92] $\square$

As a direct consequence we have $|\alpha_k| \leqslant 1$. Indeed : $\Gamma_0 \in S$ and because $\alpha_1 = \Gamma_0(0)$, $|\alpha_1| \leqslant 1$. Then

$$\Gamma_1(z) = \frac{1}{z} \frac{\Gamma_0(z) - \alpha_1}{1 - \alpha_1 \, \Gamma_1(z)} \in S \text{ and } \alpha_2 = \Gamma_1(0) \text{ thus also } |\alpha_2| \leqslant 1 \text{ etc...}$$

Thus all $|\alpha_k| \leqslant 1$. This is exactly one implication in a theorem given by Schur [1].

THEOREM 3

    $\Gamma_0(z) \in S$ iff

(1) $|\alpha_k| < 1$ $k = 1,2,\ldots$ or

(2) $|\alpha_k| < 1$ $k = 1,2,\ldots,n$ and $|\alpha_{n+1}| = 1$. In that case $\Gamma_0(z)$ is a Blaschke product of degree n and $\Gamma_n(z) = \alpha_{n+1}$ for all z and

      $\Gamma_k(z) = \alpha_{k+1} = 0$ for $k > n.$

The recursion for Γ_k or the equivalent algorithm A2 is therefore called the Schur algorithm and the α_k are sometime referred to as Schur parameters. Now we apply the converse in the above theorem for $\Pi_n(z) = Q_n^*(z)/Q_n(z)$. Recall $\Pi_{k-1}(z) = \dfrac{1}{z}\,\dfrac{\Pi_k(z) + \alpha_k}{1 + \alpha_k\,\Pi_k(z)}$ with $-\alpha_k = \Pi_k(0)$. Now all the α_k are bounded by 1 in modulus until $\alpha_0 = -1$, the previous theorem says that $\Pi_n(z)$ must be a Blaschke product of degree n and most important must be analytic in the unit disk. Thus all the zeros of $Q_n(z)$ must be in $|z| > 1$. The above recursion for $\Pi_k(z)$ is nothing else but the Schur-Cohn stability fest for $Q_n(z)$. From all this we can conclude that $c_n/Q_n(z)$ is a suitable approximation for H(z) (here $c_n = (f_0 \prod_1^n (1-\alpha_k^2))^{1/2}$).

From algorithm A1 if follows that $\hat{Q}_n(z) = Q_n^*(z)$ are the Szegö orthogonal polynomials with respect to the weight function F. I.e.

$$\frac{1}{2\Pi} \int_{-\pi}^{\pi} Q_n(e^{i\theta})\, Q_m(e^{i\theta})\, F(e^{i\theta})d\theta = \delta_{nm}\, c_n^2$$

and algorithm A1 is nothing but the recursion for these polynomials. Therefore we can also refer to the α_k as Szegö parameters. From this theory of orthogonal polynomials it follows that

$$\frac{1}{2\pi} \int_{-\pi}^{\pi} \left| 1-H(e^{i\theta})Q_n(e^{i\theta})\,\frac{1}{H(0)\,c_n}\right|^2 d\theta = \frac{1}{2\pi} \int_{-\pi}^{\pi} \left| H^{-1}(e^{i\theta})-\frac{Q_n(e^{i\theta})}{H(0)\,c_n}\right|^2 F(e^{i\theta})d\theta$$

is minimal [1,6]. Hence we have a weighted inverse least squares approximant for H.

It is clear that the parameters α_k are most important. In filtering theory they are known as <u>reflection coefficients</u>. They have a physical meaning. In the example of the speech model they are e.g. given by

$$\alpha_k = \frac{C_{k+1} - C_k}{C_{k+1} + C_k}$$

where C_k is the cross-sectional area (impedance) of cylinder k. α_k gives the fraction of energy in the sound wave that is reflected at the boundary between cylinder k and k+1 [12,13]. They are also used in the actual implementation of the filter. A possible realization is represented by the scheme

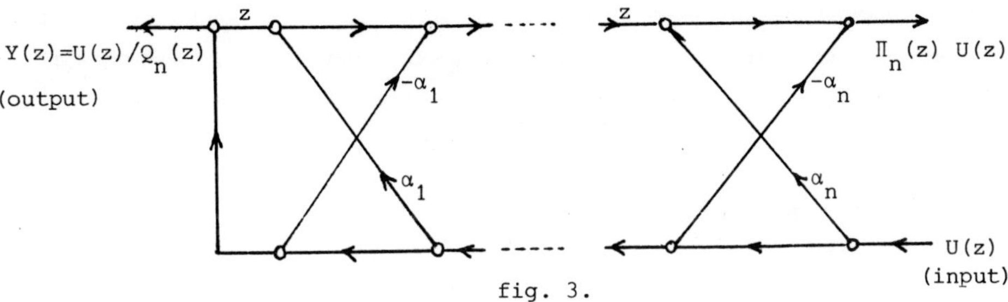

fig. 3.

It is called a lattice filter because of its structure. It is a serial connection of n elementary sections of identical type. Each section represents
a Moebius transformation which is an hyperbolic rotation. An increase in n
is also simple because we only have to add some of these elementary sections
to the scheme. The output is rather insensitive to perturbations in the α_k.
These and other reasons make the reflection coefficients most important.
They are usually computed by algorithm A1 (named afther Levinson or Durbin)
or by algorithm A2 (the Schur algorithm). A numerical stability analysis of
these algorithms was given e.g. in [3] . Algorithm A3 in this context is new.
Because of the well known numerical instability of the qd algorithm it is not
recommended for the computation of the α_k but it gives us some information
about the transmission zeros. We shall explain this in next section.

5. Transmission zeros

The zeros of H(z) are called transmission zeros. They are not only important for their physical interpretation but also because if these zeros
z_k are used as interpolation points for the Nevanlinna-Pick algorithm [1] , it
produces an interpolant for H(z) of the form $c'_n \prod_1^n (z-z_k)/Q_n(z)$ where again
the zeros of $Q_n(z)$ are outside $|z| = 1$ (a stable filter) and it is least
squares in the sense that

$$\frac{1}{2\pi} \int_{-\pi}^{\pi} \left| 1 - \frac{Q_n(e^{i\theta})}{c'_n \prod_1^n (e^{i\theta}-z_k)} \frac{H(e^{i\theta})}{H(0)} \right|^2 d\theta$$

is minimal for all possible Q_n.
If H(z) were rational and of degree n, then the AR approximant would in general never be exact for finite n. The Nevanlinna-Pick approximant however

would be of ARMA type and find an exact fit after exactly n steps if only the interpolation points are z_k, i.e. the transmission zeros of H. How does algorithm A3 now give us information about the zeros? It was shown in [4,5] that if $F(z)$ is meromorphic in $0 < |z| < \infty$, then

$$\lim_{m \to \infty} b_k^{(m)} = p_k \quad \text{and} \quad \lim_{m \to -\infty} b_k^{(m)} = p_k^{-1}$$

while $\quad \lim_{n \to \infty} a_n^{(m)} = z_{m+1} \quad , \quad m \geqslant 0$

and $\quad \lim_{n \to \infty} a_n^{(-m)} = z_m^{-1} \quad , \quad m > 0$

where $\quad 1 < |p_1| < |p_2| < \ldots \quad$ are the poles of $H(z)$

and $\quad 1 < |z_1| < |z_2| < \ldots \quad$ are the zeros of $H(z)$.

In case there are some equimodular poles or zeros, these rules can be adapted. We take e.g. the most important case of complex conjugate pairs. If we suppose that the f_k are real, then complex zeros or poles are complex conjugate. Thus suppose that

$$|z_{m-1}| < |z_m| = |z_{m+1}| < |z_{m+2}|$$

then it can be shown that

$$\lim_{n \to \infty} a_n^{(m)} a_n^{(m-1)} = z_m z_{m+1}$$

and

$$\lim_{n \to \infty} (a_{n+1}^{(m)} + a_n^{(m-1)}) = z_m + z_{m+1}.$$

From these two numbers z_m and z_{m+1} can be computed. If should be emphasized that the generation of the ab-table with the rhombus rules of algorithm A3 is unstable. Because of the relation between A1, A2 and A3 it is however possible to generate the ab-table with algorithm A1 or A2 which have a much better numerical behaviour. We need not even generate the a and b numbers explicitly because if the limits exist, then [4]

$$\lim_{n \to \infty} a_n^{(m)} = \lim_{n \to \infty} \alpha_{n-1}^{(m)} / \alpha_n^{(m)}$$

so that e.g. for m = 0, we see that the ratio of two successive reflection coefficients converges to the zero of $H(z)$ which is closest to the unit circle, provided such a zero is unique.

6. Numerical example

We give an example of how the transmission zeros are found from the computation of the $\alpha_n^{(m)}$ and $\beta_n^{(m)}$. We generated the Laurent coefficients of $F(z) = H(z) H(1/z)$ in a neighborhood of $|z| = 1$ when $H(z)$ was a real rational expression with known zeros and poles. We can use any algorithm to generate the $\alpha_n^{(m)}$ and $\beta_n^{(m)}$. Algorithm A1 and A2 give nearly the same results. Algirhtms A3 is considerably less accurate and even fails at a certain n when $H(z)$ is rational (see [4]). We give the results for algorithm A1. We can give a rule to find the point in the table where the optimal accuracy is obtained. Furtheron   rounding errors will gradually disturb the result completely. The rule is as follows : If $a_k^{(m)}$ converges to a zero with an isolated modulus, then the best value for it is found for a k such that $a_{k+1}^{(m)} = (1-\alpha_k^{(m)} \beta_k^{(m)}) \alpha_{k-1}^{(m)}/\alpha_k^{(m)} \simeq \alpha_{k-1}^{(m)}/\alpha_k^{(m)}$. I.e. when $\alpha_k^{(m)} \beta_k^{(m)}$ is almost zero. If

$$\lim_{k\to\infty} a_k^{(m)} a_k^{(m-1)} = z_m z_{m+1} \quad \text{and} \quad \lim_{k\to\infty} (a_{k+1}^{(m)} + a_k^{(m-1)}) = z_m + z_{m+1}$$

then one should stop in the $a_k^{(m)}$ row when $a_{k+1}^{(m-1)} \simeq \alpha_{k-1}^{(m-1)}/\alpha_k^{(m-1)}$ and $a_{k+1}^{(m)} \simeq \alpha_k^{(m)}/\alpha_{k+1}^{(m)}$.

Example : We considered an $H(z)$ with zeros $z_{1,2} = 2 \exp(\pm 1.5\ i)$ and $z_3 = 5$ and poles $1.5 \exp(\pm 0.75\ i)$ and $-6 \exp(\pm 0.6\ i)$. The results are :

k	$\alpha_k^{(0)}=\beta_k^{(0)}$	$\alpha_k^{(1)}$	$\alpha_k^{(2)}$	$\beta_k^{(1)}$	$\beta_k^{(2)}$
1	0.45493	0.54577	-0.22816	2.1982	1.8323
2	0.05211	2.1153	-0.45119	19.190	0.47274
3	-0.23697	-0.25124	-0.13609	-4.2199	-3.9803
4	-0.06873	-0.22721	-0.03810	-14.549	-4.4013
5	0.04618	0.07417	-0.00866	21.654	13.482
6	0.01885	0.04462	-0.00176	53.058	22.414
7	-0.01052	-0.02041	-0.00035	-95.057	-49.009
8	-0.00552	-0.00976	-0.00007	-181.23	-102.42
9	0.00223	0.00603	-0.00001	449.02	165.86

k	$a_k^{(0)}$	$\alpha_{k-2}^{(0)}/\alpha_{k-1}^{(0)}$	$a_k^{(1)}$	$\alpha_{k-1}^{(1)}/\alpha_k^{(1)}$	$a_k^{(2)}$	$\alpha_{k-2}^{(2)}/\alpha_{k-1}^{(2)}$
2	8.7064	8.7301	-10.215	-8.4195	0.61354	0.50568
5	-1.4851	-1.4883	1.8566	1.6625	4.9122	4.3986
10	1.4513	1.4513	-1.1706	-1.1692	5.0002	4.9939
11	-3.4197	-3.4197	3.7019	3.7030	5.0097	5.0112
12	1.0800	1.0800	-0.79679	-0.79697	4.9519	4.9530
13	-5.0194	-5.0194	5.3024	5.3022	5.0200	5.0199
14	0.7544	0.7544	-0.47155	-0.47153	4.9706	4.9708

k	$b_k^{(0)}$	$b_k^{(1)}$	$b_k^{(2)}$
2	0.012349	8.9263	-1.7959
5	-0.00087	-3.9354	0.19413
10	$0.6\times10(-6)$	4.8711	-0.00148
11	$-0.2\times10(-6)$	-4.4998	-0.00109
12	$0.3\times10(-7)$	6.0994	0.00018
13	$-0.9\times10(-8)$	-5.7738	0.00018
14	$0.1\times10(-8)$	9.2373	-0.00002

49

k	$\alpha_{k-2}^{(0)}\,\alpha_{k-2}^{(1)}/\alpha_{k-1}^{(0)}\,\alpha_{k-1}^{(1)}$	$\alpha_{k-2}^{(0)}/\alpha_{k-1}^{(0)} + \alpha_{k-1}^{(1)}/\alpha_{k}^{(1)}$
2	2.2524	0.31064
5	4.5590	0.17419
10	4.0033	0.28217
11	3.9982	0.28326
12	3.9994	0.28307
13	4.0003	0.28278
14	4.0000	0.28287
	$= \left\lvert z_{1,2} \right\rvert^{2}$	$= 2\,\mathrm{Re}(z_{1,2})$

References

1. Akhiezer N.I. : The classical moment problem, Oliver and Boyd, Edinburgh, London, 1965.

2. Bultheel A. : Epsilon and qd algorithms for the matrix-Padé and 2-D Padé problem. Report TW 57, 1982, K.U.Leuven.

3. Bultheel A. : Error analysis of incoming and outgoing scheme for the trigonometric moment problem. In Padé approximation and its applications, Lect. Notes Math. 888, Springer, 1981, pp. 100-109.

4. Bultheel A. : Zeros of a rational function defined by its Laurent series. Presented at the Padé conference, Bad Honnef, March 1983.

5. Bultheel A. : Quotient-difference relations in connection with AR filtering. Proc. ECCTD'83 conference, Stuttgart, Sept. 1983, VDE Verlag, Berlin, 1983.

6. Dewilde P., Dym H. : Schur recursions, error formulas and convergence of rational estimators for Stationary Stochastic sequences. IEEE Trans. info Theory, IT-$\underline{27}$ (4), 1981, pp. 446-461.

7. Gragg W.B. : The Padé table and its relation to certain algorithms of numerical analysis. SIAM Rev., $\underline{14}$, 1972, pp. 1-62.

8. Gragg W.B. : Laurent, Fourier and Chebyshev-Padé tables. In Padé and rational approximation, (Saff, Varga eds.) Ac. Press, New York, 1977, pp. 61-72.

9. Henrici P. : Applied and computational complex analysis I, Wiley, New York, 1974.

10. Jones W.B., Steinhardt A. : Digital filters and continued fractions. In Analytic theory of continued fractions, Lect. Notes Math. 932, Springer, Berlin, 1982, pp. 129-151.

11. Jones W.B., Thron W.J. : Continued fractions, Addison Wesley, London, 1980.

12. Makhoul J. : Linear prediction : a tutorial review. Proc. IEEE $\underline{63}$ (4), 1975, pp. 561-580.

13. Markel J.D., Gray A.M.Jr. : Linear prediction of speech. Springer, Berlin, 1976.

14. Rissanen N. : Algorithm for triangular decomposition of block Hankel and Toeplitz matrices, applications to factoring positive matrix polynomials, Math. Comp. $\underline{27}$, 1973, pp. 147-154.

15. Rudin W. : Real and complex analysis, McGraw-Hill, New York, 1974.

16. Thron W.J. : Two point Padé tables, T-fractions and sequences of Schur, in Padé and rational approximation. (Saff, Varga eds.). Ac. Press, New York, 1977.

Adhemar Bultheel
Katholieke Universiteit Leuven
Departement Computerwetenschappen
Celestijnenlaan 200A
3030 Heverlee (Belgium)

ISNM, Vol. 67
Numerical Methods of
Approximation Theory, Vol. 7
© 1983 Birkhäuser Verlag Basel

Einige Bemerkungen zur Numerik der multivariaten Approximation

L. Collatz

Abstract: This short paper refers to difficulties occuring in numerical treatment of multivariate Approximation, especially nonlinear Approximation (for instance for boundary value problems). Nonlinear problems become more and more important for applications, and it would be desirable to make more research in this area for which a list of open problems is given.

Inhalt: In diesem Aufsatz soll auf Schwierigkeiten hingewiesen werden, die bei der numerischen Behandlung multivariater Approximation, besonders bei nichtlinearen Problemen, auftreten; da derartige Aufgaben zu steigender Bedeutung für die Anwendungen (z.B. mit partiellen Differentialgleichungen) gelangen, soll der Aufsatz zugleich zu stärkerer Forschung auf diesem Gebiete, auf dem noch zahlreiche Fragen offen sind, anregen. Eine Anzahl wichtiger Fragestellungen wird zusammengestellt.

I. Einführung

Es soll unter Verwendung der üblichen Bezeichnungen die Tschebyscheffsche Approximation zugrundegelegt werden: In einem abgeschlossenen Bereich $\bar{B}$ des n-dimensionalen Raumes R^n der Punkte $x=(x_1,\ldots x_n)$ seien eine Funktion $f(x)\ \varepsilon\ C(\bar{B})$ und eine von Parametern $a_1,\ldots,a_p$ in linearer oder nichtlinearer Weise abhängende Familie stetiger Funktionen $W=w(x,a_1,\ldots a_p)$ gegeben, wobei die a_ν in einem Bereich A des $a_1-\ldots-a_p$-Raumes variieren können. In C werde die Halbordnung eingeführt: $v\leq z$ bedeute:

$v(x)\leq z(x)$ für alle x $\bar{B}$, wobei das Zeichen $\leq$ die klassische Ordnung reeller Zahlen bezeichnet. Alle auftretenden Größen seien reell. Die Approximation wird als Optimierung geschrieben:

$$(1.1)\qquad -\delta_1\leq w(x,a_\nu)-f(x)\leq\delta_2\text{ für alle }x\ \varepsilon\ \bar{B},$$

$$\delta_1\geq 0,\ \delta_2\geq 0,\ Q=\delta_1+\delta_2=\text{Min.}$$

Hierbei ist Q die Zielfunktion und a_ν, δ_1, δ_2 sind die Variablen. Q (oft $\frac{1}{2}Q$) wird "Minimalabstand" der Funktion f von der

$\delta_1 = 0$ (bzw. $\delta_2 = 0$) bedeutet einseitige Tschebyscheff-Approximation von oben (bzw. von unten), und $\delta_1 = \delta_2$ ist die klassische (zweiseitige) Approximation. Tritt in W ein Parameter a_j an zwei verschiedenen Stellen auf, ohne daß dies durch eine eineindeutige Transformation der Parameter vermieden werden kann, so liegt "verkettete Approximation" vor.

II. Beispiel für verkettete Approximation

Eigenwertaufgaben:

$$(2.1) \quad \begin{aligned} Lu &= \lambda Mu \text{ in } B \\ Su &= \theta \text{ auf } \partial B, \end{aligned}$$

$u(x)$ unbekannte Eigenfunktion;
L,M gegebene lineare Differentialoperatoren in B (und entsprechend S auf dem Rande ∂B); θ ist die Nullfunktion.

Als spezielles Beispiel sei genannt (etwa Eigenschwingung einer Membran oder eines Raumteiles B) (vgl. auch Fox, Henrici, Moser 1967, Kuttler Sigillito 1978 u.a.)

$$(2.2) \quad -\Delta u = -\sum_{j=1}^{n} \frac{\partial^2 u}{\partial x_j^2} = \lambda u \text{ in } B, \quad u = 0 \text{ auf } \partial B.$$

Für $u(x)$ werde eine Näherungslösung w in der Form

$$(2.3) \quad w(x, a_\nu) = \sum_{\nu=1}^{p} a_\nu w_\nu(x)$$

mit festgewählten Funktionen $w_\nu(x)$ angesetzt und der Quotient

$$(2.4) \quad \phi = \frac{Lw}{Mw} = \frac{\sum_{\nu=1}^{p} a_\nu Lw_\nu}{\sum_{\nu=1}^{p} a_\nu Mw_\nu}$$

gebildet. Unter gewissen Voraussetzungen (in einfachen Fällen vergl. etwa Collatz [63] S. 131 ff., weitgehende Verallgemeinerungen stammen von J. Albrecht [68],[83] und F. Görisch [78],[81],[83]) gilt der Einschließungssatz: Wenn für gewählte Werte der a_ν der Quotient ϕ in $\bar{B}$ zwischen endlichen Schranken

ϕ_{min} und ϕ_{max} liegt, so schließen diese Extrema mindestens einen Eigenwert λ_s von (2.1) ein. Welcher Eigenwert λ_s eingeschlossen wird, ist im allgemeinen nicht von vornherein bekannt:

$$(2.5) \qquad \phi_{min} \leq \lambda_s \leq \phi_{max}.$$

Man möchte ein möglichst kleines Einschließungsintervall I erhalten, d.h. der Quotient ϕ soll möglichst eine konstante Λ annähern; man nimmt dann Λ als Näherung für λ. Die Optimierungsaufgabe lautet dann:

$$(2.6) \qquad \left| \frac{Lw(x,y_\nu)}{Mw(x,y_\nu)} - \Lambda \right| \leq \hat{\delta} = Min$$

Wetterling [77],[81] formte dies um in

$$(2.7) \qquad \left| \frac{Lw}{Mw} - 1 \right| \leq \hat{\hat{\delta}}, \qquad \hat{\hat{\delta}} = Min.$$

Hier treten die Parameter a_ν im Zähler und im Nenner des Bruches auf. Es liegt verkettete (nichtlineare, rationale) Approximation vor. Wetterling [81] entwickelte einen Algorithmus, der im R^2 numerisch erprobt wurde; im R^3 tritt die Schwierigkeit hinzu, daß man über die Extremalstellen im allgemeinen zu wenig weiß.

III. Numerisches Beispiel

Für die Aufgabe (2.2) soll ein einfaches Beispiel als Erläuterung der Methode, ohne hohe Genauigkeitsansprüche, genannt werden; als Bereich B im R^3, wobei x,y,z statt x_1,x_2,x_3 geschrieben wird, werde gewählt

$$(3.1) \qquad B = \{(x,y,z) \text{ mit } \phi > 0, z > 0\};$$

dabei ist

$$\phi = 1-2r^2 - z^2 \text{ mit } r^2 = x^2 + y^2$$

λ ist hier proportional zu ω^2, wobei ω als Frequenz akkustischer Schwingungen im Raumteil B interpretiert werden kann;

dieser Raumteil entspricht etwa einer rotationssymmetrischen
hohen Kuppel, vgl. Fig. 1. Als Näherungsfunktion w, welche die
Randbedingung w=0 auf ∂B erfüllt, werde gewählt:

(3.2)   w=z ϕ S mit

$$S=a_1+a_2r^2+a_3z^2+a_4r^4+a_5r^2z^2+a_6z^4+P(r,z)$$

wobei das evtl. hinzukommende Polynom höhere Potenzen in r^2
und z^2 enthält.

Von den Parametern a_ν wird zunächst verlangt, daß Δw die Gestalt z ϕ h hat (etwa mit h als Polynom), damit der Quotient

$$\phi = \frac{-\Delta w}{w} = \frac{z \phi h}{z \phi S} \cdot \frac{h}{S}$$

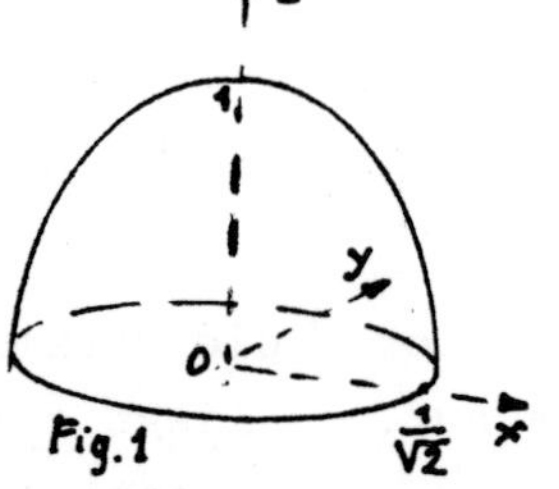

in $\bar{B}$ zwischen endlichen Schranken bleibt.

1.) Der ganz grobe Ansatz

$$a_4 = a_5 = a_6 = 0, \quad a_1 = 1:$$

würde nur $21.5 \leq \lambda_1 \leq 59.3$ liefern, man muß also mindestens
a_4, a_5, a_6 hinzunehmen. Man berechnet ein Polynom $h^{(1)}$, welches
eine möglichst gute obere Schranke liefert und ein Polynom $h^{(2)}$
für eine möglichst gute untere Schranke.

Ich danke Herrn Diplom-Mathematiker Jörg Haarmeyer für Durchführung der numerischen Rechnung auf einem Computer mit Hilfe
des Madsen-Algorithmus, implementiert von Herrn Diplom-Mathematiker Uwe Grothkopf. Man erhält:

$$h^{(1)}=1-1.101216\ r^2-1.016167z^2$$
$$\text{und } h^{(2)}=1-1.091894\ r^2-0.753667z^2$$

und die Einschließung

$$26.014 \leq \lambda_1 \leq 28.602.$$

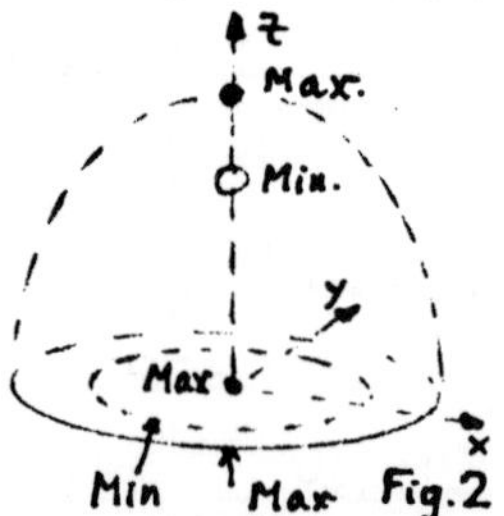

Figur 2 zeigt die Extremalstellen, die bei dem vorliegenden
rotationssymmetrischen Fall auf Kreisen (bzw. in Punkten)
liegen; im nichtrotationssymmetrischen Fall würde man eine
Anzahl diskreter Punkte als Extremalpunkte erhalten. Natürlich
kann man durch Hinzunahme weiterer Polynome in $P(r,z)$ bei S
in (3.2) die Genauigkeit leicht steigern. F. Goerisch [83] hat
in anderen Fällen gezeigt, daß man mit größeren Anzahlen von
Parametern beträchtliche Genauigkeiten erreichen kann.

IV. Numerische Schwierigkeiten der multivariaten Approximation

1. Verkettete Approximation (chained Approximation). Hierüber
gibt es nur wenige Untersuchungen; in Spezialfällen stellte
Hoffmann einen Zusammenhang mit Kontrollproblemen her. Verkettete Approximation (vgl. z.B. Collatz [73]) tritt allgemein auf,
wenn man Randwertaufgaben mit Operatoren monotoner Art oder
allgemeiner mit monoton zerlegbaren Operatoren (Collatz [68],
Bohl [74], Schröder [80] u.a.) mit Iterationsverfahren behandelt und die Ausgangsfunktionen von mehreren Parametern ab-
hängen läßt (Collatz [81], S. 194 Gl. (2.9)). Es wäre sehr erwünscht, konkret anwendbare und numerisch brauchbare Sätze
für weitere spezielle Klassen verketteter Approximationen aufzustellen, z.B. über verkettete rationale Approximation, über
Approximation, bei der nur ein Parameter a_1 an zwei oder mehr
verschiedenen Stellen auftritt u.a.

2. Aufstellung _guter_ Ausgangsnäherungen im nichtlinearen Fall.
Es ist oft schwierig, so gute Näherungen aufzustellen, daß das
Newton'sche Verfahren zur Verbesserung der Näherungen rasch
konvergiert. Die Schwierigkeiten werden häufig noch vergrößert,
wenn man mit mehreren (lokalen) Extrema zu rechnen hat.
Effektive Algorithmen gibt es bei multivariater Approximation
bisher nur in speziellen Fällen. Es fehlt hier noch viel numerische Erfahrung.

3. H-Mengen: Alternantensätze sind nur bei Funktionen einer
unabhängigen Veränderlichen anwendbar; an die Stelle der Menge
der Extremalpunkte treten bei multivariater Approximation die
H-Mengen, bzw. H_1-Mengen bei einseitiger Tschebyscheff-Approximation; sie sind für praktische Berechnungen wichtig, vgl.
auch V., weil ein Einschließungssatz (untere und obere Schranke) für den Minimalabstand, vgl. Kolmogoroff-Kriterium bei
Meinardus [67], bei Vorliegen einer H-Menge bequem angewendet
werden kann. Liegen untere und obere Schranke nahe beieinander,
so weiß man, daß man bei der betrachteten Funktionsfamilie
nahezu das Bestmögliche erreicht hat; ist die Einschließung
nicht genau genug, so muß man eine (meist umfassendere) Funk-
tionsklasse verwenden. Im nichtlinearen multivariaten Fall
wurden H-Mengen (vgl. Collatz-Krabs [73], S. 100-118) in Spezialfällen aufgestellt. Defert-Thiran [82] geben eine für die
numerische Rechnung geeignete Abänderung von H-Mengen. Es
wäre sehr erwünscht, H-Mengen in weiteren Fällen zur Verfügung
zu haben.

4. Charakterisierung der Minimallösung: Im nichtlinearen multivariaten Fall ist es nur in sehr speziellen Fällen möglich
festzustellen, ob eine ermittelte Funktion w(x) Minimallösung
ist; es gibt einige allgemeine Sätze, die sich jedoch gewöhn-
lich nicht im konkreten Fall zur Feststellung der Minimalei-
genschaft eignen; aber gerade Sätze, die für eine solche Feststellung brauchbar sind, wären wichtig.

V. Defekte

Schon bei der Tschebyscheff-Approximation einer stetigen
Funktion $f(x_1,x_2)$ durch eine lineare Funktion $w=a_1x_1+a_2x_2+a_3$
kann es eintreten, daß es nicht vier, sondern nur drei Extre-
malpunkte gibt, (in welchem der Betrag $|\varepsilon|$ des Fehlers $\varepsilon=w-f$
sein Maximum annimmt). Treten bei w die Parameter $a_1,\ldots a_p$ auf
und gibt es nur k Extremalpunkte mit $k\leq p$, so bezeichnet man
als Defekt d den Wert $d=p+1-k$; im genannten Beispiel ist $d=1$;

man beobachtet im multivariaten Fall häufig auch größere Werte von d; in manchen Fällen scheint d etwa proportional zu k zu wachsen. Größere Werte von d erschweren auch die numerische Rechnung. Es besteht daher das dringende Bedürfnis, das Phänomen der Defekte genauer zu untersuchen.

VI. Schlußbemerkung

Ebensowenig, wie man von einer allgemeinen Theorie, die <u>alle</u> nichtlinearen partiellen Differentialgleichungen umfaßt, einen Nutzen für eine spezielle Differentialgleichung erwarten kann, wird man auf eine ganz allgemeine Theorie der nichtlinearen multivariaten Approximation große Hoffnung für den konkreten Einzelfall setzen können (vgl. de Boor [82]). Wir stehen erst am Anfang einer Entwicklung, und bei der Kompliziertheit des Gebietes (vgl. de Boor [82]) sind auch kleine Erweiterungen auf weiterreichende Problemklassen schon zu begrüßen, wenn man zeigen kann, daß man damit genauere Aussagen erhalten kann. Der Prüfstein der numerischen Durchführung konkreter Beispiele ist daher äußerst wichtig, und es muß noch sehr viel mehr numerische Erfahrung gesammelt werden. Als weiteres aktuelles Anwendungsgebiet für multivariate Approximation seien auch die inversen Aufgaben genannt, vgl. Hoffmann-Komstaedt [82], für die ebenfalls noch viel Forschungsarbeit zu leisten ist.

Prof. Dr. L. Collatz
Institut für Angewandte Mathematik
der Universität Hamburg
Bundesstraße 55
2000 Hamburg 13

Literatur

Albrecht, J. [68] Verallgemeinrung eines Einschließungssatzes von
     L. Collatz, Z.Angew.Math.Mech. 48 (1968) T43-T46.

Bohl, E. [74] Monotonie, Lösbarkeit und Numerik bei Operator-
     gleichungen. Springer, 1974, 255 S.

de Boor, C. [82] Topics in Multivariate Approximation Theory,
     Springer, Lect. Notes in Math. Vol 965 (1982) 31-78.

Collatz, L. [63] Eigenwertaufgaben mit technischen Anwendungen,
     Leipzig, 1963, 500 S.

Collatz, L. [68] Funktional Analysis und Numerische Mathematik,
     Springer, 1968, 371 S.

Collatz, L. [73] Chained Approximation with Applications to
     Differential- and Integral Equations, Danish Center for
     Appl. Mech. 1973, Report Nr. 51, 14 S.

Collatz, L. [81] Anwendung von Monotoniesätzen zur Einschließung
     der Lösungen von Gleichungen, Jahrbuch Überblicke Mathem.
     1981, 189-225.

Collatz, L. - W. Krabs [73] Approximationstheorie, Teubner
     Stuttgart 1973, 2o8 S.

Defert, Ph. - J.P. Thiran [82] Exchange Algorithm for multivari-
     ate polynomials, Intern.Ser.Num.Math. 59 (1982) 115-128.

Goerisch, F. [78] Über Quotienten-Einschließungssätze bei all-
     gemeinen Eigenwertaufgaben, Intern.Ser.Num.Math. 39 (1978)
     86-1oo.

Goerisch, F. [81] Über die Anwendung einer Verallgemeinerung des
     Lehmann-Maehly-Verfahrens zur Berechnung von Eigenwert-
     schranken, Intern.Ser.Num.Math. 56 (1981) 58-72.

Hoffmann, K.H. - H.J. Kornstaedt [82] Ein numerisches Verfahren
     zur Lösung eines Identifizierungsproblems bei der Wärme-
     leitungsgleichung, Intern.Ser.Num.Math. 58 (1982) 1o8-126.

Meinardus, G. [67] Approximation of functions, Theory and numeri-
     cal methods, Springer 1967, 198p.

Schröder, J. [8o] Operator Inequalities, Acad. Press 198o, 367 S.

Wetterling, W. [77] Quotienteneinschließung bei Eigenwertaufgaben
     mit partieller Differentialgleichung, Intern.Ser.Num.Math.
     38 (1977) 213-218.

Wetterling, W. [81] Quotienteneinschließung beim ersten Membran-
     eigenwert, Intern.Ser.Num.Math. 56 (1981) 193-198.

Demnächst in Intern.Ser.Num.Math.erscheinende Arbeiten:

**Albrecht,J.[83]Einheitliche Herleitung von Einschließungssät-
     zen für Eigenwerte (Vortrag Oberwolfach Juni 1983)**

**Goerisch,F.[83] Neue Verfahren zur Berechnung von Eigenwert-
     schranken (Vortrag Oberwolfach Juni 1983)**

ISNM. Vol. 67
Numerical Methods of
Approximation Theory, Vol. 7
© 1983 Birkhäuser Verlag Basel

NUMERICAL APLICATIONS OF OPERATOR PADE APPROXIMANTS.

Annie A. M. Cuyt.

We repeat the definition of operator Padé approximant for a nonlinear operator $F: \mathbb{R}^p \to \mathbb{R}^q$; it is a generalization of the univariate Padé approximant, completely following the ideas of the univariate theory. These operator Padé approximants prove to be efficient tools for the convergence acceleration of multidimensional tables (q=1), for the solution of a system of nonlinear equations (p=q), for the numerical approximation of multivariate functions (q=1).

1. Operator Padé approximants.

Let the nonlinear operator $F: \mathbb{R}^p \to \mathbb{R}^q$ be given by its Taylor series expansion in the origin 0 in $\mathbb{R}^p$; then there exists r > 0 such that

$$F(x) = \sum_{k=o}^{\infty} \frac{1}{k!} F^{(k)}(0) x^k \qquad \text{for } \|x\| < r$$

where the $F^{(k)}(0)$ are the kth Fréchet-derivatives of F at 0 [11 pp. 110-113]

If

$$F(x) = \begin{pmatrix} F_1(x_1,\ldots,x_p) \\ F_2(x_1,\ldots,x_p) \\ \vdots \\ F_q(x_1,\ldots,x_p) \end{pmatrix}$$

then

$$\frac{1}{k!}\, F^{(k)}(0)x^k = \frac{1}{k!}\begin{pmatrix} F_1^{(k)}(0)x^k \\ \vdots \\ F_q^{(k)}(0)x^k \end{pmatrix}$$

with

$$\frac{1}{k!}\, F_\ell^{(k)}(0)x^k = \sum_{k_1+\ldots+k_p=k} \frac{1}{k_1!\ldots k_p!}\, \frac{\partial^k F_\ell}{\partial x_1^{k_1}\ldots\partial x_p^{k_p}}(0)\, x_1^{k_1}\ldots x_p^{k_p}$$

for $\ell = 1,\ldots,q$.

We denote from now on $\frac{1}{k!}F^{(k)}(0)x^k$ by $C_k x^k = \begin{pmatrix} \sum_{k_1+\ldots+k_p=k} c_{k_1\ldots k_p,1}\, x_1^{k_1}\ldots x_p^{k_p} \\ \vdots \\ \sum_{k_1+\ldots+k_p=k} c_{k_1\ldots k_p,q}\, x_1^{k_1}\ldots x_p^{k_p} \end{pmatrix}$

If we define

$$A_i x^i = \begin{pmatrix} A_{i,1}x^i \\ \vdots \\ A_{i,q}x^i \end{pmatrix}$$

$$B_j x^j = \begin{pmatrix} B_{j,1}x^j \\ \vdots \\ B_{j,q}x^j \end{pmatrix}$$

with

$$A_{i,\ell}x^i = \sum_{i_1+\ldots+i_p=i} a_{i_1\ldots i_p,\ell}\, x_1^{i_1}\ldots x_p^{i_p}$$

$$B_{j,\ell}x^j = \sum_{j_1+\ldots+j_p=j} b_{j_1\ldots j_p,\ell}\, x_1^{j_1}\ldots x_p^{j_p}$$

then it is possible [3] to calculate for n and m in $\mathbb{N}$,

$$P(x) = \sum_{i=nm}^{nm+n} A_i x^i$$

$$Q(x) = \sum_{j=nm}^{nm+m} B_j x^j$$

such that

$$(F.Q-P)(x) = \sum_{k \geqslant nm+n+m+1} D_k x^k \tag{1}$$

where the multiplication in $\mathbb{R}^q$ is performed componentwise.

In other words, such that in the power series $(F.Q-P)$ all the terms of degree less than $nm+n+m+1$ disappear.

The shift of the degrees in $P(x)$ and $Q(x)$ over nm can be motivated as follows.

Condition (1) is equivalent with the following two systems of equations.

$$(1a) \quad \begin{cases} C_o.B_{nm} \; x^{nm} = A_{nm} \; x^{nm} & \forall x \in \mathbb{R}^p \\[2ex] C_1 x.B_{nm} \; x^{nm} + C_o.B_{nm+1} \; x^{nm+1} = A_{nm+1} \; x^{nm+1} & \forall x \in \mathbb{R}^p \\[1ex] \vdots \\[1ex] C_n x^n.B_{nm} \; x^{nm} + \ldots + C_o.B_{nm+n} \; x^{nm+n} = A_{nm+n} \; x^{nm+n} & \forall x \in \mathbb{R}^p \end{cases}$$

with $B_{nm+j} \; x^{nm+j} \equiv 0$ for $j > m$

$$(1b) \quad \begin{cases} C_{n+1} \; x^{n+1}.B_{nm} \; x^{nm} + \ldots + C_{n+1-m} \; x^{n+1-m}.B_{nm+m} \; x^{nm+m} = 0 \quad \forall x \in \mathbb{R}^p \\[1ex] \vdots \\[1ex] C_{n+m} \; x^{n+m}.B_{nm} \; x^{nm} + \ldots + C_n \; x^n.B_{nm+m} \; x^{nm+m} = 0 \quad \forall x \in \mathbb{R}^p \end{cases}$$

with $C_k \; x^k \equiv 0$ for $k < o$.

A solution of (1b) can be computed by means of the following determinants in $\mathbb{R}^q$; these formulas are direct generalizations of the classical formulas for the solution of a homogeneous system :

$$B_{nm} \, x^{nm} = \begin{vmatrix} C_n \, x^n & \cdots & C_{n+1-m} \, x^{n+1-m} \\[2ex] C_{n+1} \, x^{n+1} & \cdots & C_{n+2-m} \, x^{n+2-m} \\[1ex] \vdots & & \vdots \\[1ex] C_{n+m-1} \, x^{n+m-1} & \cdots & C_n \, x^n \end{vmatrix}$$

$$B_{nm+j} \, x^{nm+j} = \begin{vmatrix} C_n \, x^n & \cdots & \boxed{-C_{n+1} \, x^{n+1}} & \cdots & C_{n+1-m} \, x^{n+1-m} \\[1ex] \vdots & & \vdots & & \vdots \\[1ex] C_{n+m-1} \, x^{n+m-1} & \cdots & \boxed{-C_{n+m} \, x^{n+m}} & \cdots & C_n \, x^n \end{vmatrix}$$

$$j^{th} \text{ column in } B_{nm} \, x^{nm} \text{ replaced by this column}$$

$$j = 1, \ldots, m$$

For every solution of (1b) a solution of (1a) can be calculated by substitution of the $B_{nm+j} \, x^{nm+j}$ $(j=0,\ldots,m)$ in the left hand side of (1a). So, using the classical formulas, we get immediately the shift of the degrees over nm in P and Q. The solution for P and Q constructed in this way need not be nontrivial. However, it is possible to prove that a nontrivial solution always exists [4], while it is not difficult to construct an operator F and choose n and m such that without the shift of the degrees in P,Q and (F.Q-P) only a trivial solution $P(x) \equiv 0 \equiv Q(x)$ can be obtained [2].

In [2] we have shown that different solutions (P_1,Q_1) and (P_2,Q_2) of the same Padé approximation problem (1) provide equivalent rational operators $\dfrac{P_1}{Q_1} \, (x)$ and $\dfrac{P_2}{Q_2} \, (x)$, where again the division in $\mathbb{R}^q$ is performed componentwise, i.e.

$$(P_1 \cdot Q_2)(x) = (P_2 \cdot Q_1)(x) \qquad \forall x \in \mathbb{R}^p$$

This conclusion enables us to define the (n,m) Operator Padé Approximant for F.

Definition 1.1. : Let the polynomials P and Q satisfy (1). We call the irreducible form $R_{n,m}(x)$ of $\frac{P}{Q}(x)$ the (n,m) Operator Padé Approximant for $F(x)$.

The $R_{n,m}$ are unique up to a multiplicative constant in numerator and denominator. To avoid this we can normalize the $R_{n,m}$, but this is not important for the applications.

For more information about these operator Padé approximants the interested reader is referred to [4] and [5].

A last remark before turning to the applications concerns the fact that for p=1=q the operator Padé approximants are nothing more than the classical univariate Padé approximants, because the shift is then merely a multiplication of numerator and denominator by x^{nm} and this is cancelled by taking the irreducible form of $\frac{P}{Q}(x)$.

2. <u>Convergence acceleration.</u>

It is well-known that the convergence of a sequence of numerical results can be accelerated by means of univariate Padé approximants. Such a sequence of results $(T_i)_{i \in \mathbb{N}}$ can be regarded as a one-dimensional table. We shall now indicate how the convergence of a p-dimensional table $(T_{i_1 \ldots i_p})_{i_1, \ldots, i_p \in \mathbb{N}}$ can be accelerated.

Before introducing the generalization, we repeat the convergence acceleration method in case p=1.

Construct

$$F(x) = \sum_{i=o}^{\infty} \Delta T_i \, x^i$$

with $\Delta T_i = T_i - T_{i-1}$ for $i \geq o$ and $T_i = 0$ for $i < o$.

Clearly

$$F(1) = \lim_{i \to \infty} T_i$$

So we can approximate

$$F(1) \simeq R_{n,m}(1)$$

and in some cases [1] accelerate the convergence of $(T_i)_{i \in \mathbb{N}}$.

In case $p > 1$, construct

$$F(x_1,\ldots,x_p) = \sum_{i_1,\ldots,i_p=o}^{\infty} \Delta T_{i_1 \ldots i_p} \; x_1^{i_1} \ldots x_p^{i_p}$$

with

$$\Delta T_{i_1 \ldots i_p} = T_{i_1 \ldots i_p} - \sum_{j=1}^{p} T_{i_1 \ldots (i_j-1) \ldots i_p}$$

$$+ \sum_{\substack{j,k=1 \\ j \neq k}}^{p} T_{i_1 \ldots i_{j-1}(i_j-1)i_{j+1} \ldots i_{k-1}(i_k-1)i_{k+1} \ldots i_p}$$

$$- \ldots + (-1)^p T_{(i_1-1) \ldots (i_p-1)}$$

It is easy to check that

$$F(1,1,\ldots,1) = \lim_{i_1,\ldots,i_p \to \infty} T_{i_1 \ldots i_p}$$

Now we can approximate F by our real-valued p-variate operator Padé approximants $R_{n,m} : \mathbb{R}^p \to \mathbb{R}$ and calculate

$$R_{n,m}(1,1,\ldots,1)$$

as an approximation for $\displaystyle\lim_{i_1,\ldots,i_p \to \infty} T_{i_1 \ldots i_p}$.

The values $R_{n,m}(1,1,\ldots,1)$ can be computed by means of the ε-algorithm [6].

We will now illustrate the procedure with some numerical results. Suppose we want to solve the following partial differential equation which is of interest in gas dynamics :

$$\Delta u(x_1,x_2) = u^2(x_1,x_2) \qquad \text{for } (x_1,x_2) \text{ in } [0,1] \times [0,1]$$
$$u(x_1,x_2) = 1 \text{ on the boundary of } [0,1] \times [0,1] \tag{2}$$

Let us discretize (2) :

choose $h_1 = \dfrac{1}{2^i}, \ h_2 = \dfrac{1}{2^j}$

define $u_{kl} = u(k\,h_1, l\,h_2)$

construct $u_{kl}^{(m)}$ in terms of $u_{kl}^{(m-1)}$ with

$$u_{k,0}^{(m)} = u_{0,1}^{(m)} = u_{k,2^j}^{(m)} = u_{2^i,1}^{(m)} = 1 \quad \text{for } k=0,\ldots,2^i \text{ and } l = 0,\ldots,2^j$$

$$u_{k,1}^{(0)} = 1$$

$$\frac{1}{h_1^2}(u_{k-1,1}^{(m)}+u_{k+1,1}^{(m)})-2\left(\frac{1}{h_1^2}+\frac{1}{h_2^2}+u_{kl}^{(m-1)}\right)u_{kl}^{(m)} +\frac{1}{h_2^2}(u_{k,1-1}^{(m)}+u_{k,1+1}^{(m)}) = -u_{kl}^{(m-1)^2}$$

for $k = 1,\ldots 2^i-1$ and $l=1,\ldots,2^j-1$

The procedure terminates when $\max\limits_{k,l}|u_{kl}^{(m)} - u_{kl}^{(m-1)}| \leqslant \epsilon$ where ϵ is the desired accuracy and this final $u_{kl}^{(m)}$ is defined to be the solution u_{kl}. To calculate a value for $u(0.5,0.5)$ for instance, we can accelerate the convergence of $T_{ij} = u(2^{i-1}h_1, 2^{j-1}h_2)$ with $h_1 = \dfrac{1}{2^i}$ and $h_2 = \dfrac{1}{2^j}$.

In this way small values for i and j would sometimes satisfy to obtain quite accurate results. In table 2.1 we compare different values of T_{ij} with different values of $R_{n,m}(1,1)$. We put $T_{io} = T_{oj} = 1$ for i=o,1,... and j=o,1,..

Table 2.1.

$T_{11} = 0.94427191$	$R_{1,1} = 1.00000000$
$T_{12} = 0.94072156$	$R_{1,2} = 0.93996831$
$T_{22} = 0.93677579$	$R_{2,2} = 0.93397294$
$T_{23} = 0.93559806$	$R_{2,3} = 0.93348171$
$T_{33} = 0.93437696$	$R_{3,3} = 0.93350234$
$T_{34} = 0.93405507$	$R_{3,4} = 0.93350688$
$T_{44} = 0.93372982$	$R_{4,4} = 0.93350940$
$T_{45} = 0.93364737$	$R_{5,4} = 0.93350936$
$T_{55} = 0.93356470$	$R_{5,5} = 0.93350934$

3. Systems of nonlinear equations.

If we want to solve a system of p nonlinear equations in p variables, we can construct iterative procedures based on the calculation of (n,m) operator Padé approximants. Let $F: \mathbb{R}^p \to \mathbb{R}^p$ be analytic in a neighbourhood of the simple root x^* that we want to compute. Since x^* is a simple root, the Jacobian $F^{(1)}(x^*)$ is a regular matrix and the inverse operator G is analytic in a neighbourhood of the origin. By F'_i and F''_i we mean respectively the first and second Fréchet-derivative [11 pp. 110] of F at $x_{(i)}$, an approximation for x^*.

The operator F'_i is represented by the Jacobian matrix

$$F'_i = \begin{pmatrix} \dfrac{\partial F_1(x)}{\partial x_1} & \cdots & \dfrac{\partial F_1(x)}{\partial x_p} \\ \dfrac{\partial F_p(x)}{\partial x_1} & \cdots & \dfrac{\partial F_p(x)}{\partial x_p} \end{pmatrix}\Bigg|_{x = x_{(i)}}$$

and the operator F_i'' by the hypermatrix

$$F_i'' = \left(\begin{array}{cc|cc|c|cc} \dfrac{\partial^2 F_1(x)}{\partial x_1^2} \cdots \dfrac{\partial^2 F_1(x)}{\partial x_1 \partial x_p} & & \dfrac{\partial^2 F_1(x)}{\partial x_2 \partial x_1} \cdots \dfrac{\partial^2 F_1(x)}{\partial x_2 \partial x_p} & & \cdots & \dfrac{\partial^2 F_1(x)}{\partial x_p \partial x_1} \cdots & \dfrac{\partial^2 F_1(x)}{\partial x_p^2} \\[4mm] \vdots & & & & & \vdots & \\[2mm] \dfrac{\partial^2 F_p(x)}{\partial x_1^2} \cdots \dfrac{\partial^2 F_p(x)}{\partial x_1 \partial x_p} & & \cdots & & & \dfrac{\partial^2 F_p(x)}{\partial x_p \partial x_1} \cdots & \dfrac{\partial^2 F_p(x)}{\partial x_p^2} \end{array} \right) \Bigg|_{x = x_{(i)}}$$

with $\dfrac{\partial^2 F_\ell(x)}{\partial x_j \partial x_k} = \dfrac{\partial^2 F_\ell(x)}{\partial x_k \partial x_j}$ for $j,k,\ell = 1,\ldots,p$.

Let $F_{(i)} = F(x_{(i)}) = y_{(i)}$ and $G(y_{(i)}) = x_{(i)}$. We know that $G(0) = x^\star$ and that G is analytic in a neighbourhood of 0; so we can write

$$G(y) = G(y_{(i)}) + G^{(1)}(y_{(i)})(y - y_{(i)}) + \frac{1}{2} G^{(2)}(y_{(i)})(y - y_{(i)})^2 + \ldots$$

$$= x_{(i)} + F_i'^{-1}(y - y_{(i)}) - \frac{1}{2} F_i'^{-1}(F_i'' F_i'^{-1})(y - y_{(i)})^2 + \ldots \tag{3}$$

where $(F_i'' F_i'^{-1})(y - y_{(i)})^2$ is the bilinear operator F_i'' evaluated on

$$(F_i'^{-1}(y - y_{(i)}), F_i'^{-1}(y - y_{(i)})).$$

If we calculate a solution of the (n,m) Padé approximation problem for G in $y_{(i)}$, we could iterate

$$x_{i+1} = R_{n,m}(0)$$

Observe that the well-known Newton – iteration results from approximating the series (3) with $y=0$ by its first two terms, i.e. a solution of the $(1,0)$ Padé approximation problem for G :

$$x_{(i+1)} = x_{(i)} - F_i'^{-1} F_{(i)} := x_{(i)} + a_{(i)}$$

The $(0,1)$ Padé approximation problem gives the following iterative method :

$$x_{(i+1)} = \frac{x_{(i)}^2}{x_{(i)} - a_{(i)}} \tag{4a}$$

The first three terms in (3) with $y=0$, which form in fact a solution for the (2,0) Padé approximation problem, could also be used to approximate x^*, giving the next iteration ;

$$x_{(i+1)} = x_{(i)} + a_{(i)} - \frac{1}{2} F_i'^{-1} F_i'' a_{(i)}^2$$

This iteration is known as Chebyshev's method for the solution of operator equations.

Another way to approximate x^* is to use a solution of the (1,1) Padé approximation problem for the series (3) :

$$x_{(i+1)} = x_{(i)} + \frac{a_{(i)}^2}{a_{(i)} + \frac{1}{2} F_i'^{-1} F_i'' a_{(i)}^2} \tag{4b}$$

which is a generalization of a formula of Frame [8] and a rediscovery of the Halley-correction, now for operator equations. The iterative procedure (4b) is closely related to the method of tangent hyperbolas [10 pp. 188]

$$x_{(i+1)} = x_{(i)} - (F_i'^{-1} + \frac{1}{2} F_i'' a_{(i)})^{-1} F_{(i)}$$

which can also be written as

$$x_{(i+1)} = x_{(i)} + (I + \frac{1}{2} F_i'^{-1} F_i'' a_{(i)})^{-1} a_{(i)}$$

where I is the identity matrix.

This second formulation shows the interrelation with (4b) : the matrix $(I + \frac{1}{2} F_i'^{-1} F_i'' a_{(i)})$ is multiplied with $a_{(i)}$, the vector $a_{(i)}$ is componentwise multiplicated with $a_{(i)}$ and those vectors are divided in order to avoid the inversion of $(I + \frac{1}{2} F_i'^{-1} F_i'' a_{(i)})$.

Let us compare the numerical effort per iterationstep for the different iterative procedures.

Iteration (4a) and Newton's method both solve one system of linear equations

$$F'_i \, a_{(i)} = -F_{(i)}$$

and combine $x_{(i)}$ and $a_{(i)}$ to find $x_{(i+1)}$.

Chebyshev's method, Halley's method and the method of tangent hyperbolas each solve two systems of linear equations

$$
\begin{aligned}
\text{Chebyshev} \quad & F'_i \, a_{(i)} = -F_{(i)} \\
& F'_i b_{(i)} = F''_i \, a_{(i)}^2 \\
& x_{(i+1)} = x_{(i)} + a_{(i)} - \tfrac{1}{2} b_{(i)}
\end{aligned}
$$

$$
\begin{aligned}
\text{Halley} \quad & F'_i \, a_{(i)} = -F_{(i)} \\
& F'_i \, b_{(i)} = F''_i \, a_{(i)}^2 \\
& x_{(i+1)} = x_{(i)} + \frac{a_{(i)}^2}{a_{(i)} + \tfrac{1}{2} b_{(i)}}
\end{aligned}
$$

$$
\begin{aligned}
\text{Tangent hyperbolas} \quad & F'_i \, a_{(i)} = -F_{(i)} \\
& (F'_i + \tfrac{1}{2} F''_i \, a_{(i)}) b_{(i)} = -F_{(i)} \\
& x_{(i+1)} = x_{(i)} + b_{(i)}
\end{aligned}
$$

However, for the first two methods these systems have the same coefficiënt matrix F'_i so that the elimination part of the Gauss-method has only to be performed once, while the third method requires the solution of linear systems with matrices F'_i and $F'_i + \tfrac{1}{2} F''_i \, a_{(i)}$ so that the entire Gauss-method has to be performed twice.

Let us now compare the numerical results for the solution of a system of nonlinear equations where the inverse operator G has singularities in the neighbourhood of 0.

Consider

$$F: \mathbb{R}^2 \to \mathbb{R}^2 : \begin{pmatrix} x_1 \\ x_2 \end{pmatrix} \to \begin{pmatrix} \exp(-x_1+x_2)-0.1 \\ \exp(-x_1-x_2)-0.1 \end{pmatrix}$$

which has a simple root

$$x^{\star} = \begin{pmatrix} -\ln 0.1 \\ 0 \end{pmatrix} = \begin{pmatrix} 2.302585092994046 \\ 0. \end{pmatrix}$$

The inverse operator

$$G: \mathbb{R}^2 \to \mathbb{R}^2 : \begin{pmatrix} y_1 \\ y_2 \end{pmatrix} \to \begin{pmatrix} -\dfrac{\ln(y_1+0.1)+\ln(y_2+0.1)}{2} \\ \dfrac{\ln(y_1+0.1)-\ln(y_2+0.1)}{2} \end{pmatrix}$$

has singularities for $y_1 = -0.1$ or $y_2 = -0.1$

In table 3.1 one finds the consecutive iterationsteps of Newton's method and iteration (4a), both of order 2, with $x_{(i)} = (5.3,0.3)$ as initial point. After 13 iterationsteps method (4a) converges $(\|x_{(13)}-x_{(12)}\| \leq 10^{-5})$ while Newton's method needs 28 iterationsteps to obtain the same accuracy. In table 3.2 one finds the results obtained by Halley's method and the method of tangent hyperbolas, both of order 3, with $x_{(o)} = (4.3,2.0)$ as initial point. If Chebyshev's method is used, starting from the same initial point $x_{(o)}$, then the sequence of iterands diverges.

Clearly methods derived from rational approximations, like Halley's method and iteration (4a), behave better in this case than methods derived from polynomial approximations, like Chebyshev's and Newton's method. The choice of the initial point also plays an important role : if it is close to the singularity, linear methods get into trouble, and if it is not, linear and rational methods can behave equally well.

Table 3.1.

	i	$x_{(i)}$	
	0	0.53000000 (+01)	0.30000000 (+00)
	1	-0.14641978 (+02)	-0.58006624 (+01)
	2	-0.13641986 (+02)	-0.58006552 (+01)
	3	-0.12642005 (+02)	-0.58006355 (+01)
	4	-0.11642059 (+02)	-0.58005821 (+01)
	5	-0.10542204 (+02)	-0.58004369 (+01)
	6	-0.96425985 (+01)	-0.58000422 (+01)
Newton	7	-0.86436705 (+01)	-0.57989703 (+01)
	8	-0.76465781 (+01)	-0.57960627 (+01)
	9	-0.56544360 (+01)	-0.57882050 (+01)
	10	-0.56754628 (+01)	-0.57671785 (+01)
	11	-0.47302660 (+01)	-0.57123764 (+01)
	12	-0.38637718 (+01)	-0.55788736 (+01)
	13	-0.31416378 (+01)	-0.53010155 (+01)
	0	0.53000000 (+01)	0.30000000 (+00)
	1	0.11128288 (+01)	0.14061045 (-01)
	2	0.29686569 (+01)	0.10780485 (-01)
	3	0.22508673 (+01)	0.36586016 (-02)
	4	0.23024184 (+01)	0.18765919 (-02)
	5	0.23025834 (+01)	0.93837391 (-03)
	6	0.23025847 (+01)	0.46918733 (-03)
(4a)	7	0.23025850 (+01)	0.23459371 (-03)
	8	0.23025851 (+01)	0.11729686 (-03)
	9	0.23025851 (+01)	0.58648482 (-04)
	10	0.23025851 (+01)	0.29324216 (-04)
	11	0.23025851 (+01)	0.14662108 (-04)
	12	0.23025851 (+01)	0.73310541 (-05)
	13	0.23025851 (+01)	0.36655270 (-05)

Table 3.2.

	i	$x_{(i)}$	
	0	.43000000000000000 (01)	.20000000000000000 (01)
	1	.3336155282457216 (01)	.1035972419924183 (01)
Halley	2	.2560818009367738 (01)	.2596797949731372 (00)
	3	.2308175634684460 (01)	.5683785304496196 (-02)
	4	.2302585151186738 (01)	.6120489087942105 (-07)
	5	.2302585092994046 (01)	-.3759322471455472 (-17)
	0	.43000000000000000 (01)	.20000000000000000 (01)
	1	.3337356399057231 (01)	.1034771307502802 (01)
Tangent Hyperbolas	2	.2561541506081360 (01)	.2589564130873139 (00)
	3	.2308222334300647 (01)	.5637241306601315 (-02)
	4	.2302585152707625 (01)	.5971357897526734 (-07)
	5	.2302585092994046 (01)	.1443269364993953 (-16)

4. Beta function.

The bivariate Beta function $B(x_1,x_2)$ is defined by

$$B(x_1,x_2) = \frac{\Gamma(x_1)\Gamma(x_2)}{\Gamma(x_1+x_2)}$$

where Γ is the gamma function

Singularities occur for $x_1 = -n$ and $x_2 = -n$ ($n = 0,1,2,\ldots$) and zeros for $x_2 = -x_1-n$ ($n = 0,1,2,\ldots$).

We write

$$B(x_1,x_2) = \frac{A(x_1-1,x_2-1)}{x_1 \cdot x_2}$$

with

$$A(u_1,u_2) = 1+u_1 u_2\ f(u_1,u_2)$$

The coefficients in the Taylor series expansion of $f(u_1,u_2)$ can be calculated by means of a method suggested in [9].

We will calculate some $R_{n,m}(u_1,u_2)$ for $f(u_1,u_2)$ (p=2) and compute

$$\frac{1+(x_1-1)(x_2-1)\ R_{n,m}(x_1-1,x_2-1)}{x_1 x_2}$$

as an approximation for $B(x_1,x_2)$. This enables us to compare the singularities and zeros of $[1+(x_1-1)(x_2-1)\ R_{n,m}(x_1-1,x_2-1)]/x_1 x_2$ with those of $B(x_1,x_2)$. The coefficients in numerator and denominator of $R_{n,m}(u_1,u_2)$ can be calculated by solving a linear system whose matrix has low displacement rank [7].

The pattern of singularities and zeros of the Beta function $B(x_1,x_2)$ itself is shown in figure 4.1 while the respective poles and zeros for $(n,m) = (0,2), (n,m) = (2,2)$ and $(n,m) = (7,1)$ are drawn in the figures 4.2 - 4.6. We remark that the vertical, horizontal and diagonal lines are nicely simulated.

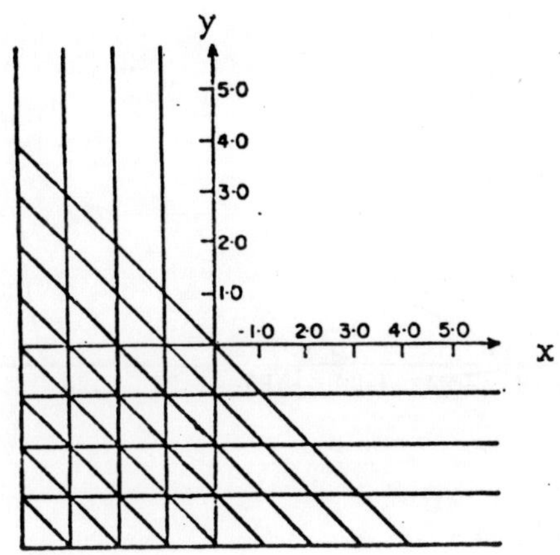

Figure 4.1.

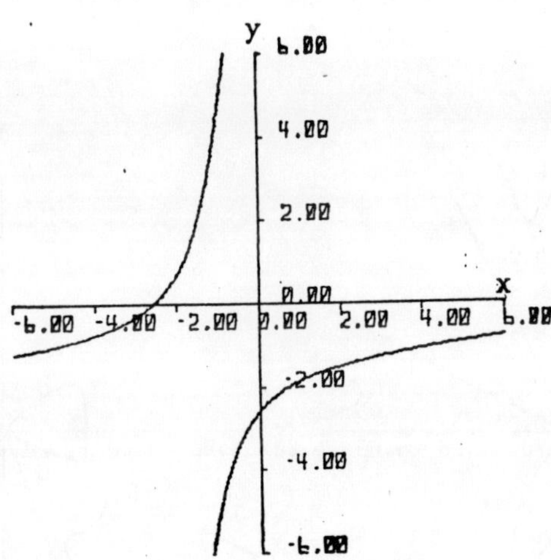

Figure 4.2.

poles

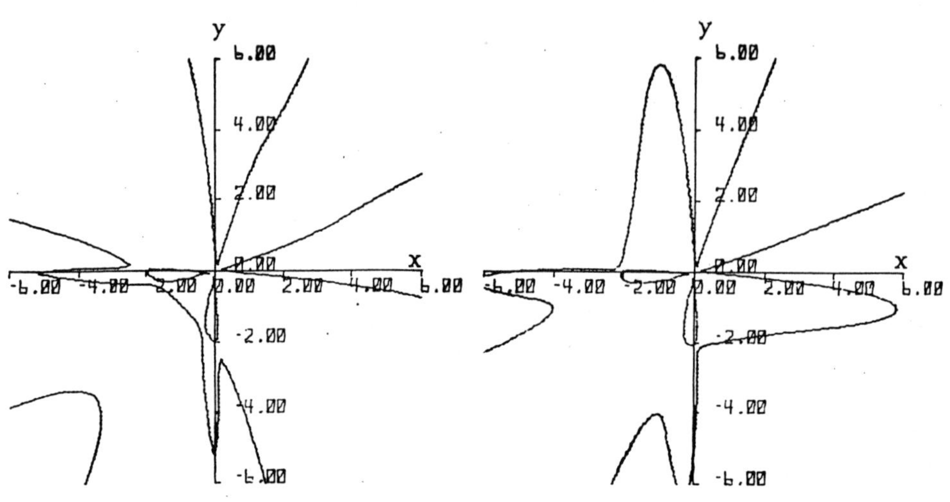

Figure 4.3.
zeros

Figure 4.4.
poles

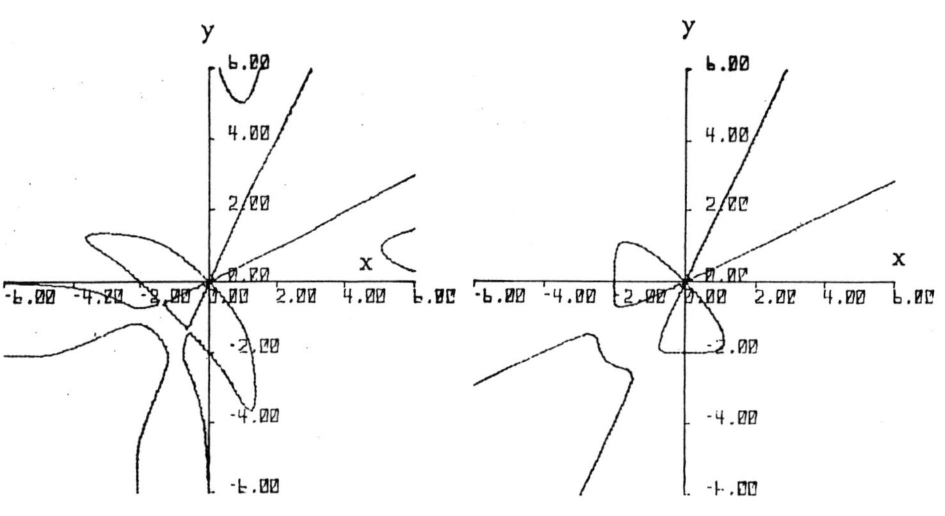

Figure 4.5.
zeros

Figure 4.6.
poles

References.

[1]   Brezinski C. Algoritmes d'accélération de la convergence. Editions Technip, Paris, 1973.

[2]   Cuyt, Annie. Abstract Padé Approximants in operator theory. LNM 765 (L. Wuytack, ed.), Springer, Berlin, 1979, pp. 61-87.

[3]   Cuyt, Annie. Regularity and normality of abstract Padé approximants Projection property and product property. Journ. Approx. Theory 35(1), 1982, pp. 1-11.

[4]   Cuyt, Annie. Multivariate Padé approximants. To appear in Journ. Math. Anal. Applics.

[5]   Cuyt, Annie. On the properties of abstract rational (1-point) approximants. Journ. Operator Th. 6(2), 1981, pp. 195-216.

[6]   Cuyt Annie. The epsilon-algorithm and multivariate Padé approximants. Num. Math. 40, 1982, pp. 39-46.

[7]   Cuyt, Annie. A comparison of some multivariate Padé approximants. Siam Journ. Math. Anal. 14(1), 1983, pp. 195-202.

[8]   Frame, J.S. The solution of equations by continued fractions. Amer. Math. Monthly 60, 1953, pp. 293-305.

[9]   Graves-Morris, P. & Hughes Jones, R. & Makinson, G. The calculation of some rational approximants in two variables. Journ. Inst. Math. Applcs. 13, 1974, pp. 311-320.

[10]   Ortega J.M. & Rheinboldt W.C. Iterative solution of nonlinear equations in several variables. Academic Press, New York and London, 1970.

[11]   Rall, L.B. Computational Solution of nonlinear operator equations, Krieger Huntington, New York, 1979;

Annie A.M. Cuyt
Research Assistant N.F.W.O.
Department of Mathematics U.I.A.
University of Antwerp
Universiteitsplein 1
B-2610   Wilrijk (Antwerp)
Belgium.

ISNM. Vol. 67
Numerical Methods of
Approximation Theory, Vol. 7
© 1983 Birkhäuser Verlag Basel

A NOTE ON RECURRENCE INTERPOLATION FORMULAE
FOR CERTAIN SETS OF POINTS IN R^k

M. Gasca and E. Lebrón

In 1960 H.C. Thacher and W.E. Milne obtained an Aitken-Neville-like recurrence interpolation formula for multivariate interpolation. In this note, we give some examples to show that the restrictive hypotheses used there can be weakened and the method can be applied in more general cases.

1. Introduction

At the moment only a few results about the application of Aitken-Neville-like recurrence formulae to multivariate interpolation problems are known. In 1960 H.C. Thacher Jr. and W.E. Milne [3] obtained a formula of this type that can only be applied to certain sets of points. In fact, the paper contains only one example, the simplicial lattice, given by the set

$$S = \left\{ (x_i, y_j) \in R^2 \mid i,j = 0,1,\ldots,n \; ; \; i+j \leq n \right\} \quad (1)$$

where $x_i \neq x_h$, $y_j \neq y_k$ if $i \neq h$, and $j \neq k$.

The interpolation problem of finding $p \in P_n$ (space of polynomials in x, y with total degree non greater than n) such that

$$p(X) = f(X) \quad \forall X \in S \quad (2)$$

where f is a given function, has a unique solution that is obtained from the solutions of three similar problems, with n replaced by n-1, respectively associated to the sets

$$S_1 = \left\{ (x_i, y_j) \mid i,j=0,1,\ldots,n-1; \; i+j \le n-1 \right\}$$

$$S_2 = \left\{ (x_i, y_j) \mid i=1,2,\ldots,n \; ; \; j=0,1\ldots,n-1 \; ; \; i+j \le n \right\} \qquad (3)$$

$$S_3 = \left\{ (x_i, y_j) \mid i=0,1,\ldots,n-1; \; j=1,2,\ldots,n \; ; \; i+j \le n \right\}.$$

More generally, the solution of the problem (2) of order n in R^k, $k \ge 2$, is constructed from the solutions of k+1 problems of order n-1.

The aim of this note is to show that the hypotheses given in [3] can be weakened and the idea can be applied to solve interpolation problems with more general sets of points.

First, we recall the hypotheses in [3] . Let S be a set with

$$W_n (k) = \binom{n+k}{k}$$

points of R^k, such that the interpolation problem (2) has a unique solution in P_n , space of polynomials in $x^{(1)}$, $x^{(2)},\ldots,$ $x^{(k)}$ with total degree non greater than n.

Let us assume the existence of k+1 subsets S_j, (j=1,2, $\ldots$,k+1) of S verifying the following conditions:

i) $$S = \bigcup_{j=1}^{k+1} S_j$$

ii) For each j=1,2,$\ldots$,k+1 the interpolation problem (2), with S replaced by S_j , has a unique solution in P_{n-1}. (Note that this implies card. $S_j = W_{n-1}(k)$).

iii) For each j=1,2,$\ldots$,k+1 there exists a unique point $X_j \in S_j$ such that $X_j \notin S_h$ $\forall h \ne j$, and if we write

$$T = \left\{ X_1, X_2, \ldots, X_{k+1} \right\} \quad ,$$

then the interpolation problem (2) associated with T has a unique solution in P_1. We denote by $X_{k+2}, X_{k+3},\ldots,X_{W_n(k)}$ the remaining points of S.

iv) For each $j > k+1$ and each $h = 1, 2, \ldots, k+1$ if $X_j \notin S_h$ then the interpolation problem (2) associated with the set

$$\{x_j\} \cup \bigcup_{\substack{t=1 \\ t \neq h}}^{k+1} X_t$$

is not unisolvent in P_1.

Under these conditions, the solution of the problem (2) can be written in terms of the solutions $p_{n-1}^{(i)}$, $i = 1, 2, \ldots, k+1$ of the problems (2) associated with S_i (and P_{n-1}) by

$$p_n = \frac{\begin{vmatrix} p_{n-1}^{(1)} & x_1^{(1)} - x^{(1)} & \ldots\ldots & x_1^{(k)} - x^{(k)} \\ p_{n-1}^{(2)} & x_2^{(1)} - x^{(1)} & \ldots\ldots & x_2^{(k)} - x^{(k)} \\ \cdot & \cdot & & \cdot \\ \cdot & \cdot & & \cdot \\ \cdot & \cdot & & \cdot \\ p_{n-1}^{(k+1)} & x_{k+1}^{(1)} - x^{(1)} & & x_{k+1}^{(k)} - x^{(k)} \end{vmatrix}}{\begin{vmatrix} 1 & x_1^{(1)} & \ldots\ldots & x_1^{(k)} \\ 1 & x_2^{(1)} & \ldots\ldots & x_2^{(k)} \\ \cdot & \cdot & & \cdot \\ \cdot & \cdot & & \cdot \\ \cdot & \cdot & & \cdot \\ 1 & x_{k+1}^{(1)} & \ldots\ldots & x_{k+1}^{(k)} \end{vmatrix}} \tag{5}$$

where $(x_j^{(1)}, x_j^{(2)}, \ldots, x_j^{(k)}) = X_j$

For $k = 1$ this is the well-known formula of the Aitken-Neville method.

Obviously, the example (1)-(3), with $X_1 = (x_0, y_0)$, $X_2 = (x_n, y_0)$, $X_3 = (x_0, y_n)$, satisfies all the conditions above. Formula (5) can be applied repeatedly to provide an algorithm for the construction of p_n.

<u>2. Some remarks about Thacher-Milne´s assumptions.</u>

We emphasize that (5) can be applied to sets of points different from simplicial lattices. For example, the lattice in figure 1 has been called a "natural lattice" by Chung and Yao 1 .

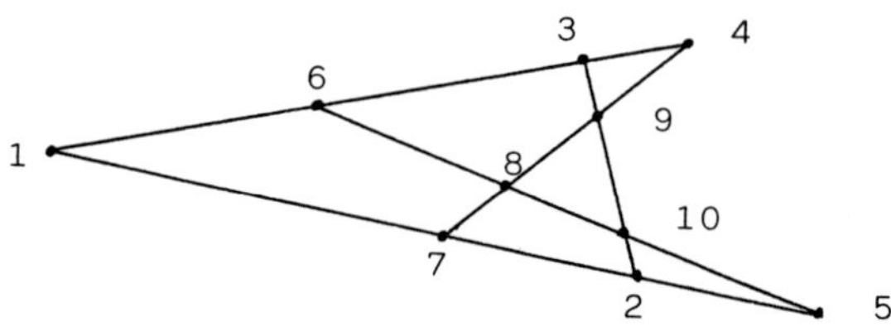

Figure 1.

For this lattice we can take

$$S = \left\{ 1,2,3,4,\ldots\ldots,10 \right\}$$
$$S_1 = \left\{ 1,4,5,6,7,8 \right\}$$
$$S_2 = \left\{ 2,5,7,8,9,10 \right\}$$
$$S_3 = \left\{ 3,4,6,8,9,10 \right\}.$$

Secondly, let us consider a rectangular lattice of $(m+1)\times(n+1)$ points in R^2:

$$S = \left\{ (x_i, y_j) \mid i=0,1,\ldots,m; \ j=0,1,\ldots,n \right\}. \tag{6}$$

The interpolation problem (2) is now unisolvent in the space $Q_{m,n}$ of polynomials of degree non greater than m in x and non greater than n in y.

Let S_j (j=1,2,3,4) be the following sets:

$$S_1 = \left\{ (x_i, y_j) \mid i=0,1,\ldots,m-1; \ j=0,1,\ldots,n-1 \right\}$$
$$S_2 = \left\{ (x_i, y_j) \mid i=1,2,\ldots,m \ ; \ j=0,1,\ldots,n-1 \right\}$$
$$S_3 = \left\{ (x_i, y_j) \mid i=1,2,\ldots,m \ ; \ j=1,2,\ldots,n \right\} \tag{7}$$
$$S_4 = \left\{ (x_i, y_j) \mid i=0,1,\ldots,m-1; \ j=1,2,\ldots,n \right\}$$

which satisfy

$$S = \bigcup_{h=1}^{4} S_h .$$

If we take $X_1 = (x_0, y_0)$, $X_2 = (x_m, y_0)$, $X_3 = (x_m, y_n)$, $X_4 = (x_0, y_n)$, then similar conditions to ii) iii) are verified, because the interpolation problem (2) is unisolvent in $Q_{m-1,n-1}$ when associated with S_h and in $Q_{1,1}$ when associated with

$$T = \left\{ X_1, X_2, X_3, X_4 \right\} .$$

Finally, condition iv) is fulfilled with 4 (resp. $Q_{1,1}$) instead of k+1 (resp. P_1).

All that leads us to think about a formula similar to (5):

$$p_{m,n} = \frac{\begin{vmatrix} p_{m-1,n-1}^{(1)} & x_0-x & y_0-y & x_0y_0-xy \\ p_{m-1,n-1}^{(2)} & x_m-x & y_0-y & x_my_0-xy \\ p_{m-1,n-1}^{(3)} & x_m-x & y_n-y & x_my_n-xy \\ p_{m-1,n-1}^{(4)} & x_0-x & y_n-y & x_0y_n-xy \end{vmatrix}}{\begin{vmatrix} 1 & x_0 & y_0 & x_0y_0 \\ 1 & x_m & y_0 & x_my_0 \\ 1 & x_m & y_n & x_my_n \\ 1 & x_0 & y_n & x_0y_n \end{vmatrix}} , \qquad (8)$$

where $p_{m-1,n-1}^{(h)} \in Q_{m-1,n-1}$ is the solution of the interpolation problem (2) for S_h and $Q_{m-1,n-1}$.

In order to show that these conjectures are really confirmed, we denote by q the right hand-side of (8), which belongs to $Q_{m,n}$. We must prove that

$$q(x_i, y_j) = f(x_i, y_j) \qquad \forall (x_i, y_j) \in S.$$

If $(x_i, y_j) \in S_h \quad \forall h$, then

$$p_{m-1, n-1}^{(h)}(x_i, y_j) = f(x_i, y_j) \qquad \forall h$$

and some simple calculations show that

$$q(x_i, y_j) = f(x_i, y_j) \ .$$

Let us now consider, for example, the case

$$(x_i, y_j) \in S_h, \qquad h=1, \ 2 \tag{9}$$

$$(x_i, y_j) \notin S_h, \qquad h=3, \ 4. \tag{10}$$

Then $q(x_i, y_j)$ can be written

$$q(x_i, y_j) = \frac{\begin{vmatrix} f(x_i, y_j) & x_0 - x_i & y_0 - y_j & x_0 y_0 - x_i y_j \\ f(x_i, y_j) & x_m - x_i & y_0 - y_j & x_m y_0 - x_i y_j \\ p_{m-1, n-1}^{(3)}(x_i, y_j) & x_m - x_i & y_n - y_j & x_m y_n - x_i y_j \\ p_{m-1, n-1}^{(4)}(x_i, y_j) & x_0 - x_i & y_n - y_j & x_0 y_n - x_i y_j \end{vmatrix}}{\begin{vmatrix} 1 & x_0 & y_0 & x_0 y_0 \\ 1 & x_m & y_0 & x_m y_0 \\ 1 & x_m & y_n & x_m y_n \\ 1 & x_0 & y_n & x_0 y_n \end{vmatrix}} =$$

$$\frac{\begin{vmatrix} f(x_i,y_j) & x_0-x_i & y_0-y_j & x_0y_0-x_iy_j \\ 0 & x_m-x_0 & 0 & x_my_0-x_0y_0 \\ p^{(3)}_{m-1,n-1}(x_i,y_j) - f(x_i,y_j) & x_m-x_0 & y_n-y_0 & x_my_n-x_0y_0 \\ p^{(4)}_{m-1,n-1}(x_i,y_j) - f(x_i,y_j) & 0 & y_n-y_0 & x_0y_n-x_0y_0 \end{vmatrix}}{\begin{vmatrix} 1 & x_0 & y_0 & x_0y_0 \\ 1 & x_m & y_0 & x_my_0 \\ 1 & x_m & y_n & x_my_n \\ 1 & x_0 & y_n & x_0y_n \end{vmatrix}} \; .$$

The coefficient of $f(x_i,y_j)$, in the expansion of the numerator by its first column, is

$$\begin{vmatrix} x_m-x_0 & 0 & x_my_0-x_0y_0 \\ x_m-x_0 & y_n-y_0 & x_my_n-x_0y_0 \\ 0 & y_n-y_0 & x_0y_n-x_0y_0 \end{vmatrix} = \begin{vmatrix} 1 & x_0 & y_0 & x_0y_0 \\ 1 & x_m & y_0 & x_my_0 \\ 1 & x_m & y_n & x_my_n \\ 1 & x_0 & y_n & x_0y_n \end{vmatrix}$$

The coefficient of $p^{(3)}_{m-1,n-1}(x_i,y_j)-f(x_i,y_j)$ is

$$\begin{vmatrix} x_0-x_i & y_0-y_j & x_0y_0-x_iy_i \\ x_m-x_0 & 0 & x_my_0-x_0y_0 \\ 0 & y_n-y_0 & x_0y_n-x_0y_0 \end{vmatrix}$$

which is zero, because (9) (10) imply that $y_j=y_0$, and then the first two rows are linearly dependent.

The coefficient of $p^{(4)}_{m-1,n-1}(x_i,y_j) - f(x_i,y_j)$ also vanishes by analogous reasons, so that finally

$$q(x_i, y_j) = f(x_i, y_j).$$

The other cases for (x_i, y_j) can be treated similarly.

A generalization of formula (8) can easily be found for R^k, in the form of a quotient of two determinants of order 2^k.

It is useful to point out that the set S (6) can also be written in a more convenient way as the union of

$$S_1 = \left\{ (x_i, y_j) \mid i=0,1,\ldots,m; j=0,1,\ldots,n-1 \right\}$$
$$S_2 = \left\{ (x_i, y_j) \mid i=0,1,\ldots,m; j=1,2,\ldots,n \right\}.$$

The problem (2), for S_h $(h=1,2)$, is unisolvent in $Q_{m-1,n-1}$, but there is not a <u>unique</u> point $X_1 \notin S_2$ (resp. $X_2 \notin S_1$).

If we choose, for example, $X_1 = (x_0, y_0)$, $X_2 = (x_0, y_n)$, then the problem (2), for $T = \{X_1, X_2\}$, has a unique solution in $Q_{0,1}$, and the hypothesis iv) in section 1 is fulfilled with 2 (resp. $Q_{0,1}$) instead of $k+1$ (resp. P_1).

It is easily seen that in this case

$$p_{m,n} = \frac{\begin{vmatrix} p_{m,n-1}^{(1)} & y_0 - y \\ p_{m,n-1}^{(2)} & y_n - y \end{vmatrix}}{\begin{vmatrix} 1 & y_0 \\ 1 & y_n \end{vmatrix}} \qquad (11)$$

where $p_{m,n-1}^{(h)}$ is the solution of the problem (2) for S_h and $Q_{m,n-1}$. A similar formulation can be written for R^k in the form of a quotient of two determinants of order two.

Finally, let us observe that there are no changes in (11) if, for example, the data

$$P_{m,n}(x_3,y) = f(x_3,y_j) \qquad \forall j$$

are replaced by

$$\frac{\partial\, P_{m,n}(x_2,y_j)}{\partial x} = \frac{\partial f(x_2,y_j)}{\partial x} \qquad \forall j \ .$$

Therefore, formulae of types considered here can also be used for some Hermite interpolation problems.

References.

|1| Chung, K.C. & Yao, T.H.: On lattices admitting unique Lagrange interpolation. SIAM J. Num. Anal. 14 (1977) 735-743.

|2| Gasca, M. & López-Carmona, A.: A general recurrence interpolation formula and its applications to multivariate interpolation. J. of Approx. Theo. 34 (1982) 361-374.

|3| Thacher Jr. H.C.& Milne, W.E. : Interpolation in several variables. J. SIAM (1960) 33-42.

M. Gasca
Depto. Ecuaciones Funcionales
Facultad Ciencias (Matemáticas)
Universidad de Zaragoza (Spain)

E. Lebrón
Depto. Ecuaciones Funcionales
Facultad de Ciencias
Universidad de Granada (Spain)

ISNM, Vol. 67
Numerical Methods of
Approximation Theory, Vol. 7
© 1983 Birkhäuser Verlag Basel

REAL VS. COMPLEX RATIONAL CHEBYSHEV APPROXIMATION
ON COMPLEX DOMAINS

Martin H. Gutknecht   and   Lloyd N. Trefethen*

Let S be a Jordan region symmetric about the real axis, and consider best maximum norm approximation on S of an analytic (or merely continuous) function f satisfying $f(\bar{z}) = \overline{f(z)}$ by a rational function of type (m,n) with either real or complex coefficients. For $m = 0$ and $n \geq 4$, the error in complex approximation can be arbitrarily much smaller than the error in real approximation. In contrast, for $(m,n) = (0,1)$ the complex error can be better by at most a constant factor.

1. Introduction ; statement of the result

Let $S \subset \mathbb{C}$ be a compact point set symmetric about the real axis, let C^r be the set of complex functions f defined and continuous on S and satisfying $f(\bar{z}) = \overline{f(z)}$, let $A^r \subset C^r$ be the subset of functions analytic in the interior of S , let $E^C_{mn}(f)$ be the error in best approximation (with respect to the maximum norm) of f on S from the set R^C_{mn} of complex rational functions with numerator degree at most m and denominator degree

---

* Supported in part by a National Science Foundation Mathematical Sciences Postdoctoral Fellowship.

at most n , and let $E^r_{mn}(f)$ denote the corresponding error for
the subset $R^r_{mn} \subset R^c_{mn}$ with real coefficients. Finally, set

$$\gamma^S_{mn} := \inf_{f \in A^r \setminus R^r_{mn}} \frac{E^c_{mn}(f)}{E^r_{mn}(f)} .$$

For the case where $S = I$ is an <u>interval</u> of the real
line (and hence $A^r = C^r$), A.A. Gončar seems to have been the
first to notice that for some $f \in C^r$, complex approximations
are better than real ones. From the work of K.N. Lungu,
E.B. Saff and R.S. Varga [4, 6], and A. Ruttan [3] various conditions (necessary or sufficient) are known for $E^c_{mn}(f) < E^r_{mn}(f)$,
and in particular, it is known that $\gamma^I_{mn} < 1$ for all $m \geq 0$,
$n \geq 1$. However, no triple (f,m,n) with $E^c_{mn}(f) < \frac{1}{2} E^r_{mn}(f)$
and no positive lower bound for any γ^I_{mn} have been known until
recently, when we established the results

$$\gamma^I_{mn} = 0 \quad \text{for} \quad n \geq m+3 , \tag{1}$$

$$\gamma^I_{01} > 0 \tag{2}$$

[2, 5]. At the same time we showed that for the closed <u>unit
disk</u> Δ

$$\gamma^\Delta_{0n} = 0 \quad \text{for} \quad n \geq 4 \tag{1'}$$

[2, 5], and we mentioned that our arguments for (1') and (2) can
be extended to the case where S is the closure of a Jordan
region symmetric about $\mathbb{R}$ whose boundary is differentiable at
the two points of intersection with the real axis:

$$\gamma^S_{0n} = 0 \quad \text{for} \quad n \geq 4 , \tag{1''}$$

$$\gamma^S_{01} > 0 . \tag{2''}$$

The proof of (1") is short, and it is described in [5]. In contrast, the proof of (2") is tedious, and the similar proof of (2)
was only outlined in [5]. Here, we now want to state the whole

proof for the case of a symmetric Jordan region. Analyticity of
f turns out to be inessential, i.e. we prove a slightly more
general result:

THEOREM. Let $S \subset \mathbb{C}$ be the closure of a Jordan region
which is symmetric about the real axis and whose boundary is
differentiable at the two points of intersection with the real
axis. Then

$$\inf_{f \in C^r \setminus R^r_{01}} \frac{E^c_{01}(f)}{E^r_{01}(f)} > 0 . \tag{3}$$

A fortiori, $\gamma^S_{01} > 0$.

We do not claim that (1), (1'), and (1") have much
significance for practical approximation problems. In a real-
istic application, $E^c_{mn}(f)$ will rarely be much less than $E^r_{mn}(f)$,
and even if it is, an increase in degree of a real approximation
will probably be more cost-effective than a switch to complex
coefficients. In fact, if $f \in C^r$, than for any $c \in R^c_{mn}$ the
associated function $r(z) := \frac{1}{2}[c(z) + \overline{c(\overline{z})}]$ lies in $R^r_{m+n,2n}$
and

$$\|f - r\| = \frac{1}{2} \max_{z \in S} \left| [f(z) - c(z)] + [f(z) - \overline{c}(z)] \right| \leq \|f - c\| .$$

In particular, $E^r_{m+n,2n}(f) \leq E^c_{mn}(f)$, cf. [4, Prop. 1].

2. Proof

Let $f \in C^r$ be arbitrary, and let c^* be a best
approximation of f from R^c_{mn} . For any $c \in R^c_{mn}$ set

$$\overline{c}(z) := \overline{c(\overline{z})} , \qquad \hat{c}(z) := \frac{1}{2}[c(z) - \overline{c}(z)] .$$

Then

$$\|\hat{c}^*\| = \frac{1}{2} \max_{z \in S} \left| [f(z) - c^*(z)] - [f(z) - \overline{c}^*(z)] \right|$$

$$\leq \; \|f - c*\| \; = \; E_{mn}^{c}(f) \; .$$

Since $E_{mn}^{r}(f) \leq E_{mn}^{c}(f) + \|c* - r\|$ for every $r \in R_{mn}^{r}$, it follows that

$$E_{mn}^{r}(f) \; \leq \; E_{mn}^{c}(f) \; + \; \|\hat{c}*\| \; \frac{\|c* - r\|}{\|\hat{c}*\|} \; \leq \; E_{mn}^{c}(f) \left[1 + \frac{\|c* - r\|}{\|\hat{c}*\|} \right] \qquad (4)$$

if $\|\hat{c}*\| \neq 0$, i.e. $c* \notin R_{mn}^{r}$.

Now suppose that for any $c \in R_{mn}^{c} \smallsetminus R_{mn}^{r}$ with no poles on S we can find $r_c \in R_{mn}^{r}$ (depending on c) such that

$$\frac{\|\hat{c}\|}{\|c - r_c\|} \; \geq \; \delta \qquad (5)$$

for some fixed $\delta > 0$. According to (4) this implies

$$\frac{E_{mn}^{c}(f)}{E_{mn}^{r}(f)} \; \geq \; \frac{1}{1 + 1/\delta} \qquad (6)$$

whenever $c* \notin R_{mn}^{r}$. But otherwise $E_{mn}^{c}(f) = E_{mn}^{r}(f)$ trivially. Therefore if (5) holds, (6) is true for all $f \in C^{r} \smallsetminus R_{mn}^{r}$.

Our proof for the case $(m,n) = (0,1)$ consists in defining a suitable mapping $c \mapsto r_c$ and in verifying (5) for this mapping.

First, without loss of generality we may assume that $\pm 1 \in \partial S$. Then, as in [5], $S^{c} := \overline{\mathbb{C}} \smallsetminus S$ (where $\overline{\mathbb{C}} := \mathbb{C} \cup \{\infty\}$) is split up into

$$A^{\pm} \; := \; \{z \in \mathbb{C} \; ; \; |\arg(-1 \pm z)| < \theta\} \cup \{\infty\} \; ,$$

$$C \quad := \; \{z \in \mathbb{C} \; ; \; |z| \geq \rho\} \smallsetminus (A^{+} \cup A^{-}) \; ,$$

$$B \quad := \; \mathbb{C} \smallsetminus (A^{+} \cup A^{-} \cup C \cup S) \; ,$$

as indicated in Fig. 1. $\theta \in (0, \pi/4)$ and ρ are assumed to be chosen such that

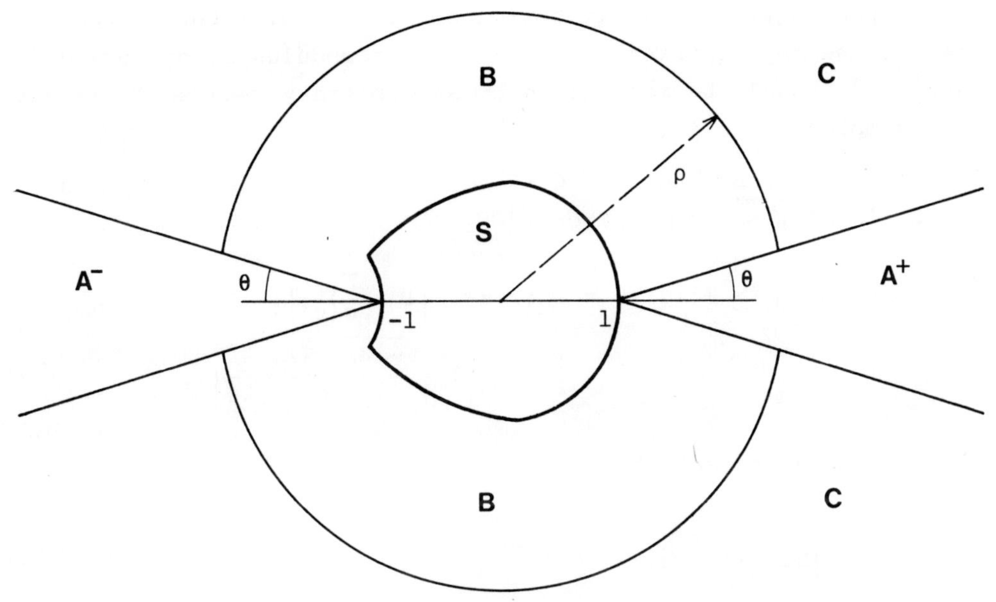

<u>Fig. 1</u>

$$2S \cap C = \emptyset, \quad \text{i.e.} \quad \frac{\rho}{2} > \rho_S := \max\{|z|; \ z \in S\}, \tag{7}$$

$$S \cap \{z \in \mathbb{C}; \ |\arg(-1+\sigma z)| < 2\theta\} = \emptyset, \quad \sigma = \pm 1, \tag{8}$$

$$\rho \geq 4/\sin\theta. \tag{9}$$

For any $c \in R_{01}^C \setminus R_{00}^C$,

$$c(z) := \frac{a}{1 - z/z_0},$$

we define r_c depending on the position of the pole z_0 by

$$r_c(z) := \begin{cases} \dfrac{1 - 1/|z_0|}{1 \mp z/|z_0|} \ \text{Re} \ c(\pm 1) & \text{if} \quad z_0 \in A^{\pm}, \\[2mm] 0 & \text{if} \quad z_0 \in B, \\[2mm] \text{Re} \ a & \text{if} \quad z_0 \in C. \end{cases} \tag{10}$$

Correspondingly, our verification of (5) is split into three cases. We may restrict a to have unit modulus since c, $\hat{c}$, and r_c can all be multiplied by an arbitrary real scale factor. The simplest case is

$\underline{z_0 \in B}$. Let $z^* \in S$ be a point closest to z_0, so that $|c(z^*)| = \|c\|$. Then

$$\|\hat{c}\| \geq |\hat{c}(z^*)| = \frac{1}{2} |c(z^*)| \left| 1 - \frac{\overline{c}(z^*)}{c(z^*)} \right|$$

$$= \frac{1}{2} \|c\| \left| 1 - \frac{\overline{a}}{a} \frac{\overline{z}_0}{z_0} \frac{z_0 - z^*}{\overline{z}_0 - z^*} \right| . \tag{11}$$

Due to the differentiability of ∂S at ± 1,

$$\left| \frac{z_0 - z^*}{\overline{z}_0 - z^*} \right| \leq \cos \theta < 1 \tag{12}$$

whenever $z_0 \in B$ is close enough to ± 1, say, $z_0 \in U_B^{\pm}$. On the other hand, since $B \smallsetminus (U_B^{+} \cup U_B^{-})$ has a positive distance from the real axis, there is a fixed $\gamma_B < 1$ such that $|z_0 - z^*| / |\overline{z}_0 - z^*| \leq \gamma_B$ for all z_0 in this set. Therefore, in view of (10)–(12) we get for all $z_0 \in B$

$$\|\hat{c}\| \geq \|c - r_c\| \delta_B \quad \text{with} \quad \delta_B := \frac{1}{2} \min\{1 - \cos \theta, \ 1 - \gamma_B\}. \tag{13}$$

$\underline{z_0 \in C}$. In view of

$$c(z) = a + \frac{az}{z_0} [1 + h(z)] \quad \text{with} \quad h(z) := \frac{z/z_0}{1 - z/z_0} ,$$

we get

$$\|\hat{c}\| \geq \max_{z = \pm 1} |\hat{c}(z)| = \max_{z = \pm 1} |\operatorname{Im} c(z)|$$

$$= \max_{\pm} \left| \operatorname{Im}\left[a \pm \frac{a}{z_0} \left(1 + h(\pm 1) \right) \right] \right| .$$

Now, by (9), $|z_0| \geq \rho \geq 4/\sin \theta \geq 4$ for $z_0 \in C$, hence

$$|h(\pm 1)| \leq \frac{1}{|z_0|(1-|z_0|^{-1})} \leq \frac{2}{|z_0|} \leq \frac{2}{\rho} \leq \frac{1}{2}\sin\theta \qquad (15)$$

A fortiori,

$$\|\hat{c}\| \geq |\operatorname{Im} a| - \frac{2}{|z_0|} \, . \qquad (16)$$

Likewise, using (7) we get

$$\|c - r_c\| = \max_{z\in S}\left| i\operatorname{Im} a + \frac{az}{z_0}\left[1 + \frac{z}{z_0} + \left(\frac{z}{z_0}\right)^2 + \ldots\right]\right|$$

$$\leq |\operatorname{Im} a| + 2\,\frac{\rho_S}{|z_0|} \, . \qquad (17)$$

From (16) and (17) we conclude that for $\operatorname{Im} a \neq 0$ and $3/|\operatorname{Im} a| \leq |z_0| \leq \infty$

$$\frac{\|\hat{c}\|}{\|c - r_c\|} \geq \frac{|\operatorname{Im} a| - 2/|z_0|}{|\operatorname{Im} a| + 2\rho_S/|z_0|} \geq \frac{1}{3 + 2\rho_S} \, . \qquad (18)$$

For $\operatorname{Im} a \neq 0$ and $\rho \leq |z_0| \leq 3/|\operatorname{Im} a|$ inequality (17) implies $\|c - r_c\| \leq (3 + 2\rho_S)/|z_0|$, and therefore using (14) we get for $|z_0|$ in this range

$$\frac{\|\hat{c}\|}{\|c - r_c\|} \geq \frac{1}{3 + 2\rho_S} \max_{\pm}\left| \operatorname{Im}\left[a|z_0| \pm \frac{a\overline{z_0}}{|z_0|}\left(1 + h(\pm 1)\right)\right]\right|$$
$$(19)$$

Now for $z_0 \in C$ $(z_0 \neq \infty)$ the two disks

$$D^{\pm} := \left\{ a|z_0| \pm \frac{a\overline{z_0}}{|z_0|}(1 + \zeta) \; ; \; |\zeta| \leq \frac{1}{2}\sin\theta \right\}$$

lie on opposite sides of the ray $a\,\mathbb{R}^+$ at a distance of at least $\frac{1}{2}\sin\theta$ from it, and also outside the disk $|z| \leq 2$, cf. Fig. 2. Therefore, at least one of them has a distance $\geq \frac{1}{2}\sin\theta$ from the real axis, and (18) and (19) imply

$$\frac{\|\hat{c}\|}{\|c - r_c\|} \geq \frac{\sin\theta}{6 + 4\rho_S} \quad \text{if} \quad z_0 \in C \quad \text{and} \quad \operatorname{Im} a \neq 0 \, . \qquad (20)$$

By continuity, the same bound also holds if $\operatorname{Im} a = 0$. (Note

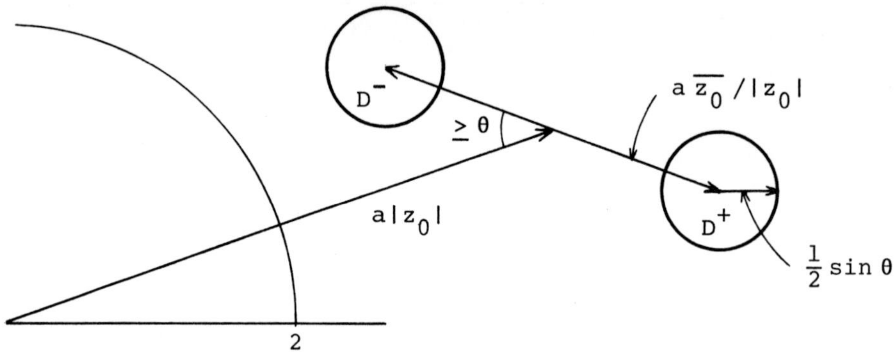

<u>Fig. 2</u>

that $\hat{c} \neq 0$ for $z_0 \in C$.)

$z_0 \in A^+$. Here we again use the estimate $\|\hat{c}\| \geq$ max$\{ |\hat{c}(1)|, |\hat{c}(-1)| \}$; now the case where $\hat{c}(1) = \text{Im } c(1) = 0$, i.e. $a = \pm a_0$ with

$$a_0 := \frac{1 - 1/z_0}{|1 - 1/z_0|} , \qquad (21)$$

will require special attention. In view of the invariance with respect to real factors of a mentioned above, we may restrict ourselves to the plus sign and set

$$a = a_0 \, e^{i\alpha} \quad (|\alpha| \leq \pi/2), \qquad d := \tfrac{1}{2}(e^{-2i\alpha} - 1),$$

so that $\bar{a} = \overline{a_0} \, e^{-i\alpha} = \overline{a_0} \, (1 + 2d) \, e^{i\alpha}$ and

$$\text{Re } c(1) = \text{Re } \frac{e^{i\alpha} a_0}{1 - 1/z_0} = \frac{\cos \alpha}{|1 - 1/z_0|} = \frac{e^{i\alpha}(1 + d)}{|1 - 1/z_0|} . \qquad (22)$$

Then

$$\hat{c}(z) = \frac{1}{2} \frac{e^{i\alpha}}{|1 - 1/z_0|} \left[\frac{z_0 - 1}{z_0 - z} - (1 + 2d)\frac{\overline{z_0} - 1}{\overline{z_0} - z} \right] ,$$

or,

$$\hat{c}(z) = \frac{e^{i\alpha}|z_0|}{|z_0 - 1|} \cdot \frac{i(1 - z)\,\text{Im}\,z_0 - d(z_0 - z)(\overline{z_0} - 1)}{(z_0 - z)(\overline{z_0} - z)} \,, \quad (23)$$

and by (22),

$$c(z) - r_c(z) = c(z) - \frac{|z_0| - 1}{|z_0| - z}\,\text{Re}\,c(1)$$

$$= \frac{e^{i\alpha}}{|1 - 1/z_0|}\left[\frac{z_0 - 1}{z_0 - z} - (1 + d)\frac{|z_0| - 1}{|z_0| - z}\right]$$

$$= \frac{e^{i\alpha}|z_0|}{|z_0 - 1|} \cdot \frac{(z_0 - |z_0|)(1 - z) - d(z_0 - z)(|z_0| - 1)}{(z_0 - z)(|z_0| - z)} \,.$$

$$(24)$$

In particular,

$$|\hat{c}(1)| = \frac{|z_0||d|}{|z_0 - 1|} \,. \quad (25)$$

Now for $z_0 > 1$ (24) yields

$$\|c - r_c\| = \frac{z_0\,|d|}{|z_0 - 1|}\,\sup_{z \in S}\,\frac{z_0 - 1}{|z_0 - z|} \leq \frac{z_0\,|d|}{|z_0 - 1|}\cdot\frac{1}{\sin 2\theta}\,,$$

and, since $d \neq 0$ if $c \notin R_{mn}^r$, we get, using (25),

$$\frac{\|\hat{c}\|}{\|c - r_c\|} \geq \sin 2\theta \quad \text{if} \quad z_0 > 1\,. \quad (26)$$

Likewise, using additionally (8) and the relation

$$z_0 - |z_0| = \frac{i\,e^{i\phi/2}}{\cos(\phi/2)}\,\text{Im}\,z_0\,, \quad \text{where} \quad \phi := \arg(z_0 - 1)\,, \quad (27)$$

we obtain asymptotically for $|\text{Im}\,z_0| = \beta|d| \to 0$ with fixed $\beta > 0$

$$\frac{\|\hat{c}\|}{\|c - r_c\|} \geq \sin 2\theta\left[1 + \frac{\beta(1 + \rho)}{\sin\theta\,\cos(\theta/2)\,(|z_0| - 1)^2}\right]^{-1} + o(1)\,. \quad (28)$$

On the other hand, if $d = 0$, we get from (23) and (24)

$$\frac{\|\hat{c}\|}{\|c - r_c\|} \geq \inf_{z \in S} \left| \frac{\hat{c}(z)}{c(z) - r_c(z)} \right| = \frac{|\operatorname{Im} z_0|}{|z_0 - |z_0||} \inf_{z \in S} \left| \frac{|z_0| - z}{\overline{z_0} - z} \right|$$

$$\geq \frac{1}{2} \cos \frac{\theta}{2} \quad \text{if} \quad z_0 \in A^+, \quad \alpha = d = 0 \,. \tag{29}$$

Now, clearly, $\|\hat{c}\| > 0$ unless $\alpha = 0$ and $\operatorname{Im} z_0 = 0$, which means that $c = r_c \in R_{01}^r$. Conversely, $c = r_c$ implies $\alpha = 0$, $\operatorname{Im} z_0 = 0$, $\|\hat{c}\| = 0$. Hence $\|\hat{c}\| / \|c - r_c\|$ is a positive continuous function of

$$(z_0, \alpha) \in \left((\overline{A^+} \smallsetminus \{1, \infty\}) \times [-\tfrac{\pi}{2}, \tfrac{\pi}{2}] \right) \smallsetminus \mathbb{R} \times \{0\} \,. \tag{30}$$

Since $\overline{A^+} \times [-\pi/2, \pi/2]$ is a compact subset of $\overline{\mathbb{C}} \times \mathbb{R}$, it suffices to establish positive lower bounds for $\|\hat{c}\| / \|c - r_c\|$ that hold in the three cases

$$(z_0, \alpha) \to \mathbb{R} \times \{0\}, \quad z_0 \to 1, \quad z_0 \to \infty \,.$$

But for $1 < |z_0| < \infty$ the first case has just been treated, and (26), (28), (29) are appropriate bounds, except that z_0 should not approach 1 in (28). Hence we are left with the two other cases and may assume $z_0 \notin \mathbb{R}$:

For $z_0 \to 1$ and $\phi = \arg(z_0 - 1) \neq 0$, one can deduce from (23), (24), and (27) that for fixed $z \neq 1$ asymptotically

$$|\hat{c}(z)| \sim \frac{|i \sin \phi - d e^{-i\phi}|}{|1 - z|}, \quad |c(z) - r_c(z)| \sim \frac{\left| 2 i \sin \frac{\phi}{2} e^{i\phi/2} - d \right|}{|1 - z|} \,.$$

Hence $\|\hat{c}\| \sim |\hat{c}(1)|$, $\|c - r_c\| \sim |c(1) - r_c(1)|$; but $\hat{c}(1) = c(1) - r_c(1) \; (\forall z_0)$ as is easily checked, and therefore

$$\frac{\|\hat{c}\|}{\|c - r_c\|} \to 1 \,. \tag{31}$$

Finally, for $z_0 \to \infty$, (23) and (24) imply

$$|\hat{c}(z)| \to |d|, \qquad |c(z) - r_c(z)| \to |d| \qquad (32)$$

uniformly for all $z \in S$. Hence (31) holds again if $d \neq 0$. The case $d = 0$ is covered by (29).

This concludes the proof. □

Except for the case $z_0 \in B$ this proof remains valid if S is replaced by an interval $I \subset \mathbb{R}$. In fact, in the two other cases we have used very few properties of S except its symmetry, namely (7), (8), and $\pm 1 \in \partial S$.

References
----

1. Ellacott, S.W.: A note on a problem of Saff and Varga concerning the degree of complex rational approximation to real valued functions. Bull. Amer. Math. Soc. *6* (1982), 218-220.

2. Gutknecht, M.H., Trefethen, L.N.: Real and complex Chebyshev approximation on the unit disk and interval. Bull. Amer. Math. Soc. *8* (1983), 455-458.

3. Ruttan, A.: The length of the alternation set as a factor in determining when a best real rational approximation is also a best complex rational approximation. J. Approx. Theory *31* (1981), 230-243.

4. Saff, E.B., Varga, R.S.: Nonuniqueness of best complex rational approximations to real functions on real intervals. J. Approx. Theory *23* (1978), 78-85.

5. Trefethen L.N., Gutknecht, M.H.: Real vs. complex rational Chebyshev approximation on an interval. Trans. Amer. Math. Soc. (1983, in print).

6. Varga, R.S.: Topics in Polynomial and Rational Interpolation and Approximation. Montréal, Les Presse de l'Université de Montréal, 1982.

Martin H. Gutknecht
Seminar für Angew. Mathematik
Eidg. Techn. Hochschule
   Zürich
ETH-Zentrum
CH-8092 Zürich, SCHWEIZ

Lloyd N. Trefethen
Courant Institute of
   Mathematical Sciences
New York University
251 Mercer Street
New York, N.Y. 10012, USA

ISNM, Vol. 67
Numerical Methods of
Approximation Theory, Vol. 7
© 1983 Birkhäuser Verlag Basel

CONVERGENCE OF THE SEQUENCE OF ITERATES OF NONEXPANSIVE MAPPINGS (A SURVEY)

R. Guzzardi, S.P. Singh & B. Watson

In this note a brief survey, covering a very limited aspect, on the convergence of the sequence of iterates is given.

The sequence of successive approximations for nonexpansive mappings, unlike contraction mappings, may fail to converge. For example, if

$$f:R \to R \text{ is given by } f(x) = 1 - x,$$

then $x_{n+1} = fx_n$ gives, for $x_o = 1$ say, $x_{2n} = 1$ and $x_{2n+1} = 0$ for $n \geq 1$. Also, rotation about the origin in the plane is another example where $x_{n+1} = fx_n$ $(x_o \neq 0)$ does not converge.

An early result, concerning the convergence of the sequence of successive approximations, is due to Krasnoselskii [15]. In 1955 he proved the following.

THEOREM 1. Let X be a uniformly convex Banach space and C a closed, convex subset of X. If $f:C \to C$ is nonexpansive (that is,

$$\| f(x) - f(y) \| \leq \| x - y \| \quad \text{for any } x, y \in C)$$

and $\overline{f(C)}$ is compact, then the mapping defined by

$$f_{\frac{1}{2}}x = \tfrac{1}{2}x + \tfrac{1}{2}fx$$

has the property that its sequence of iterates always converges to a fixed point of f.

Since f and $f_{\frac{1}{2}}$ have the same fixed points, the limit of a convergent sequence given by

$$x_{n+1} = \tfrac{1}{2}x_n + \tfrac{1}{2}fx_n$$

is necessarily a fixed point of f.

More generally if C is a convex set in a Banach space X and $f:C \to C$ is a nonexpansive mapping, then for $\lambda \in (0,1)$,

---

Supported in part by NSERC and NATO grants.

$$f_\lambda x = \lambda x + (1 - \lambda)fx$$

is a nonexpansive map and has the same fixed points as f. Schaefer [21] proved that the sequence $x_{n+1} = \lambda x_n + (1 - \lambda)fx_n$ converges to a fixed point of f under the assumption of Theorem 1. Unlike a contraction mapping, a non-expansive mapping has more than one fixed point. In this case it turns out that the limit of $x_{n+1} = \lambda x_n + (1 - \lambda)fx_n$ can depend on x_o and on λ as well.

In 1966 Edelstein [9] succeeded in relaxing the condition of uni-form convexity and proved Theorem 1 for strictly convex Banach spaces. This was a significant improvement. Diaz and Metcalf [6] gave a theorem for strictly convex Banach spaces for sequences of the type $x_{n+1} = \lambda x_n + (1 - \lambda)fx_n$.

Recall the following definitions. Let C be a bounded subset of a Banach space X. The measure of noncompactness, $\alpha(C)$, of C is defined by $\alpha(C) = \inf\{\epsilon > 0 \,|\, C$ admits a finite covering consisting of subsets of diameter $\leq \epsilon\}$. Let $f:X \to X$ be a continuous mapping. Then f is said to be <u>densifying</u> if for all bounded sets $A \subset X$, with f(A) bounded, $\alpha(f(A)) < \alpha(A)$ for $\alpha(A) > 0$.

In 1971 Petryshyn [16] extended the result to densifying non-expansive mappings. Recall that a nonexpansive mapping f with $\overline{f(C)}$ compact is a special case of a densifying mapping.

The condition that f be densifying (or, in particular that $\overline{f(C)}$ be compact) can not be eliminated. Genel and Lindenstrauss [12] have shown that there exists a closed, bounded, convex subset C of ℓ^2, a Hilbert space, and a nonexpansive mapping $f:C \to C$ such that for some $x_o \in C$, the sequence $\{x_n\}$ defined by

$$x_{n+1} = \tfrac{1}{2}(x_n + fx_n), \quad n = 0, 1, \ldots$$

has no convergent subsequence. In 1973 some further results in this direction were given by Petryshyn and Williamson Jr. [17]. They considered a quasi-nonexpansive mapping defined as follows. Let $f:C \to X$ and let F(f) denote the set of fixed points of f. Then f is called quasi-nonexpansive if $F(f) \neq \phi$ and

$$\| fx - fp \| \leq \| x - p \| \quad \text{for all } x \in C, \text{ and } p \in F(f).$$

A nonexpansive mapping f with $F(f) \neq \phi$ is always quasi-nonexpansive, but the converse is not true. A quasi-nonexpansive map need not be continuous. Consider the following example due to Dotson Jr. [7]. Let $f:R \to R$ be defined by

$$f(x) = \frac{x}{2} \sin \frac{1}{x} \quad \text{if } x \neq 0 \text{ and } f(0) = 0.$$

Then $F(f) \neq \phi$ and f is quasi-nonexpansive. This follows easily since $F(f) = \{0\}$. If $x \neq 0$, then $f(x) \neq x$, because $f(x) = x = \frac{x}{2} \sin \frac{1}{x}$ would give $2 = \sin \frac{1}{x}$ which is impossible. But, f is not nonexpansive as $\| fx - fy \| \leq \| x - y \|$ is not satisfied for $x = \frac{2}{9\pi}$ and $y = \frac{3}{19\pi}$.

In 1970 Dotson Jr. [7] considered the iteration process given below and discussed the convergence of the sequence given by $x_{n+1} = (1 - c_n)x_n + c_n fx_n$. Let C be a closed subset of a Banach space S and let $f:C \to X$. Let $\{c_n\}$ be a sequence such that $c_n \in (0,1]$ for each n. Let $x_1 \in C$ be such that x_{n+1} is defined by

$$x_{n+1} = (1 - c_n)x_n + c_n fx_n$$

for each n. If $0 < c_n < 1$ and Σc_n diverges then $\{x_n\}$ is a normal Mann process [7].

Results in the same direction are given by Reinermann [19] and Rhoades [20]. In a recent more important development Edelstein and O'Brien [11], and Ishikawa [14] independently proved that even strict convexity is not essential. Edelstein and O'Brien considered $f_\lambda x = \lambda fx + (1 - \lambda)x$, $(f_\lambda^n(x_o) = x_n)$ whereas Ishikawa considered the sequence of the type

$$x_{n+1} = (1 - c_n)x_n + c_n fx_n,$$

where $\{c_n\} \in (0,1)$, $c_n \leq \ell < 1$ and $\Sigma c_n = +\infty$.

They independently proved that the sequence defined above has the property that

$$\lim_{n \to \infty} \| x_{n+1} - x_n \| = 0.$$

If the range of f is precompact, then the sequences $\{f_\lambda^n x_o\}$, and $\{x_n\}$ where $x_{n+1} = (1 - c_n)x_n + c_n fx_n$ converge to a fixed point of f.

The following theorem is due to Petryshyn and Williamson Jr. [17].

THEOREM 2. Let C be a closed subset of a Banach space X and let $f:C \to X$ be continuous such that

  i) $F(f) \neq \phi$,

  ii) for each $x \in C$ and every $p \in F(f)$

$$\| fx - p \| \leq \| x - p \| , \text{ and}$$

  iii) there exists $x_o \in C$ such that $x_n = f^n x_o \in C$ for each $n \geq 1$.

Then $\{x_n\}$ converges to a fixed point of f in C if and only if $\lim_{x \to \infty} d(x_n, F(f)) = 0$.

In 1981, Das, Singh & Watson [5] gave the following result.

THEOREM 3. Let C be a closed subset of a Banach space X and let $f:C \to X$ be quasi-nonexpansive with $F = F(f)$. Suppose that $x_1 \in C$ is such that

$$x_{n+1} = c_n fx_n + (1 - c_n)x_n$$

yields $\{x_n\}$ either as a sequence in a normal Mann process or a sequence of iterates $(x_{n+1} = fx_n)$. If $\lim\limits_{n \to \infty} d(x_n, F) = 0$, then $\{x_n\}$ converges to a fixed point of f.

Remarks:

i) In this case, $\Sigma\, c_n$ divergent is not required. Also, it is evident that if $c_n = 1$ for each n, then the theorem holds for complete metric spaces.

ii) If $c_n = 1$ for each n, one gets Theorem 1.2 of [17] in so far as f is not assumed continuous.

S. Reich [18] proved the following.

THEOREM 4. Let C be a closed convex subset of a uniformly convex Banach space with Fréchet differentiable norm, and suppose $f:C \to C$ is nonexpansive and has a fixed point. Let $\{c_n\}$ be a sequence of real numbers satisfying $0 \le c_n \le 1$ and $\Sigma\, c_n(1 - c_n) = +\infty$. If $x_1 \in C$ and if $\{x_n\}$ is defined by

$$x_{n+1} = (1 - c_n)x_n + c_n fx_n \quad \text{for } n \ge 1$$

then $\{x_n\}$ converges weakly to a fixed point of f.

The following interesting result is due to Edelstein [10].

THEOREM 5. If (1) X is a strictly convex Banach space (2) $f:X \to X$ is a non-expansive mapping such that for some $x_o \in X$ and a sequence of integers $n_1 < n_2 < n_3 < \ldots$, we have

$$f^{n_i}x_o \to \bar{x}_o \, ,$$

and (3) $x_n = \dfrac{1}{n} \sum\limits_{i=1}^{n} f^i x_o$ has a subsequence $\{x_{n_i}\}$ converging weakly to $\bar{x} \in X$,

then   i) $\bar{x}$ is a fixed point, and

ii) $x_n \to \bar{x}$.

The following result, known as the Mean Ergodic Theorem for non-linear contractions has been given.

THEOREM 6. Let C be a closed, bounded, convex subset of a uniformly convex Banach space with Fréchet differentiable norm, and let $f:C \to C$ be non-expansive. For $x \in C$, let

$$s_n(x) = \frac{1}{n} \sum_{i=0}^{n-1} f^i(x).$$

Then $s_n(x)$ converges weakly to a fixed point of f.

The above theorem was proved in Hilbert space by Baillon [1] and then he extended to L^p spaces [2]. Bruck [3] gave a much simpler proof of the present theorem. In a recent paper Bruck [4] has shown that

$$\| s_n(x) - fs_n(x) \| \to 0 \text{ as } n \to \infty$$

uniformly for $x \in C$, when the only assumption is that X is uniformly convex.

The following is due to Dotson Jr. [8].

THEOREM 7. Let H be a Hilbert space and $f:H \to H$ a monotonic, nonexpansive operator on H. For $y_o \in H$, define $T:H \to H$ by $Tu = -fu + y_o$ for all $u \in H$. Suppose $0 \le c_n \le 1$ for all $n = 1, 2, \ldots$, and $\sum_1^\infty c_n(1 - c_n)$ diverges. Then

$$x_{n+1} = (1 - c_n)x_n + c_n Tx_n$$

converges to the unique solution $u = p$ of the equation

$$u + fu = y_o.$$

Goebel and Kirk [13] very recently established a metric inequality in a setting which includes all normed linear spaces and all spaces with hyperbolic metric, and this inequality is then applied to the study of the iteration process for approximating fixed points of nonexpansive mappings. A general theorem is obtained which unifies both the Ishikawa [14], and the Edelstein and O'Brien [11] generalizations of Krasnoselskii's method [15] of successive approximations.

References

1. Baillon, J.B. Un théoréme de type ergodique pour les contractions non linéaires dans un espace de Hilbert. C.R. Acad. Sci. Paris 280 (1975) 1511-1514.

2. Baillon, J.B. Comportement asymptotique des itérés de contractions non linéaires dans les espaces L^p, C.R. Acad. Sci. Paris 286 (1978) 157-159.

3. Bruck, R.E. A simple proof of the mean ergodic theorem for nonlinear contractions in Banach spaces, Israel J. Math. 32 (1979) 297-282.

4. Bruck, R.E. On the convex approximation property and the asymptotic behaviour of nonlinear contractions in Banach spaces, Israel J. Math. 38 (1981) 304-314.

5. Das, K.M., Singh, S.P. & Watson, B. A note on Mann Iteration for Quasi-nonexpansive Mappings, Nonlinear Analysis, 6 (1981) 675-676.

6. Diaz, J.B. & Metcalf, F.T. On the set of subsequential limit points of successive approximations, Trans. Amer. Math. Soc. 135 (1969) 459-485.

7. Dotson, Jr., W.G. On the Mann Iterative Process, Trans. Amer. Math. Soc. 149 (1970) 65-73.

8. Dotson, Jr., W.G. An iterative process for nonlinear monotonic non-expansive operators in Hilbert Space, Math. Comp. 32 (1978) 223-225.

9. Edelstein, M. A remark on a theorem of M.A. Krasnoselskii, Amer. Math. Monthly 73 (1966) 509-510.

10. Edelstein, M. An Nonexpansive Mappings of Banach spaces, Proc. Cambridge Philos. Soc. 60 (1964) 439-447.

11. Edelstein, M. & O'Brien, R.C. Nonexpansive Mappings, Asymptotic Regularity and Successive Approximations, J. London Math. Soc. (2) 17 (1978) 547-554.

12. Genel, A. and Lindenstrauss, J. An example concerning fixed points, Israel J. Math. 22 (1975) 81-86.

13. Goebel, K. & Kirk, W.A. Iteration Process for nonexpansive mappings, Proc. of the Conf. Fixed Point Theory & Applications, Amer. Math. Soc., (to appear).

14. Ishikawa, S. Fixed Points and Iteration of a nonexpansive mappings in a Banach space, Proc. Amer. Math. Soc. 59 (1976) 65-71.

15. Krasnoselskii, M.A. Two remarks on the method of successive approx-imations, Uspehi Mat. Nauk. 10 (1955) 123-127.

16. Petryshyn, W.V. Structure of fixed point sets of k-set contractions, Arch. Rat. Mech. Anal. 40 (1971) 312-238.

17. Petryshyn, W.V. & Williamson Jr., T.E. Strong and Weak Convergence of the sequence of successive approximations for quasinonexpansive mappings, J. Math. Anal. & Appl. 43 (1973) 459-497.

18. Reich, S. Weak convergence theorems for nonexpansive mappings in Banach spaces. J. Math. Anal. Appl. 67 (1979) 274-276.

19. Reinermann, J. UeberToeplitzsche Iterationsverfahren und einige ihrer Anwendungen in der Konstructiven Fixpunkttheorie, Studia Math. 32 (1969) 209-227.

20. Rhoades, B.E. Fixed Point Iterations using Infinite Matrices, Trans. Amer. Math. Soc. 196 (1974) 161-176.

21. Schaefer, H. Über die Methode Sukzessiven Approximationen, Jahre. Deutsch Math. Verein 59 (1957) 131-140.

DIPARTIMENTO DI MATEMATICA
UNIVERSITA DELLA CALABRIA
87036 ARCAVACATA DI RENDE (CS)
ITALY

DEPARTMENT OF MATHEMATICS
MEMORIAL UNIVERSITY
ST. JOHN'S, NEWFOUNDLAND
CANADA A1B 3X7

ISNM. Vol. 67
Numerical Methods of
Approximation Theory, Vol. 7
© 1983 Birkhäuser Verlag Basel

AN APPROXIMATION PROBLEM IN NAVIGATION

David Handscomb

The material presented here represents only the author's initial thoughts on what would be a good and practical way to formulate the given problem, prior to his embarking on serious research.

The context of the problem is as follows. A moving object sets out from a known origin $\underline{a}$ (in $\mathbb{R}^2$ or $\mathbb{R}^3$) to travel through a fluid medium (air or water). Its velocity $\underline{v}(t)$ relative to the medium is known (in magnitude and direction) at all times, but the velocity $\underline{v}_m(\underline{x},t)$ of the medium itself is unknown except possibly in general terms. The only other information available consists of intermittent measurements $f(t_i)$ (i=1,2,3,...) of the local value of a physical quantity $g(\underline{x})$, which varies from place to place but not from time to time, whose variations have previously been recorded on a map. From these data we wish to estimate the actual locus of the object.

An example is furnished by a ship navigating in fog, when its course and speed through the sea are known but the strength of the ocean currents are not, assuming that the only other clue to its position is given by soundings of the depth of the seabed immediately below the ship. Without this further information, the only possible estimate of position is that given by 'dead-reckoning',

$$\underline{x}(t) = \underline{a} + \int_0^t \underline{v}(s)\, ds,$$

entirely ignoring the unknown effects of ocean currents. If this is seriously in error, the fact will become apparent when one examines the discrepancies

$$e_i = f(t_i) - g(\underline{x}(t_i));$$

the question is how one should make best use of these discrepancies to improve on the dead reckoning.

The measurements $f(t_i)$ are of course subject to (considerable) experimental error, and the function $g(\underline{x})$ also (which is probably the result of applying an interpolation process to a large set of similar measurements) incorporates some error, but suppose for the present that both are known exactly. Then we are looking for a locus $\underline{x}(t)$ such that $e_i = 0$ $(i=1,2,3,\ldots)$, $\underline{x}(0) = \underline{a}$. If we now make the (dubious) assumption that the ocean current at any point has a Gaussian distribution with constant spread, and is independent of the current at any other point, a possible strategy is to seek to minimize

$$\int_0^T \underline{v}_m(\underline{x},t)^2\, dt = \int_0^T \{\underline{x}'(t) - \underline{v}(t)\}^2\, dt,$$

given that $\underline{x}(0) = \underline{a}$ and that $\underline{x}(t_i)$ lies on the contour

$$g(\underline{x}(t_i)) = f(t_i),$$

for $i=1,2,3,\ldots$.

Straightforward application of the calculus of variations shows that any solution to this problem has the property that $\underline{x}'(t)-\underline{v}(t)$ is piecewise constant, with jump discontinuities at the points t_i only, while the vector representing the discontinuity at t_i is parallel to the gradient $\nabla g(\underline{x}(t_i))$ (if this is meaningful) and, finally, $\underline{x}'(t) = \underline{v}(t)$ when $T > t >$ the last t_i (t_n, say). We have therefore to solve the following system of nonlinear equations for $\underline{x}_i = \underline{x}(t_i)$ $(i=1,2,\ldots,n,$ while $\underline{x}_0 = \underline{a})$ and $\underline{v}_{mi} = \{\underline{x}'(t) - \underline{v}(t), t_{i-1} < t < t_i\}$:

$$\begin{cases} \underline{x}_i - \underline{x}_{i-1} - (t_i-t_{i-1})\,\underline{v}_{mi} = \displaystyle\int_{t_{i-1}}^{t_i} \underline{v}(s)\, ds \quad (i=1,2,\ldots,n), \\[2ex] g(\underline{x}_i) = f(t_i) \quad (i=1,2,\ldots,n), \\[2ex] \underline{v}_{m(i+1)} - \underline{v}_{mi} \parallel \nabla g(\underline{x}_i) \quad (i=1,2,\ldots,n-1), \\[2ex] \underline{v}_{mn} \parallel \nabla g(\underline{x}_n). \end{cases}$$

An obvious way to proceed is to make the approximation

$$g(\underline{x}_i) \cong g(\underline{X}_i) + (\underline{x}_i - \underline{X}_i)\cdot \nabla g(\underline{X}_i),$$

where $\underline{X}_i$ is a previous estimate (possibly the dead-reckoning estimate)
of $\underline{x}_i$, and
$$\nabla g(\underline{x}_i) \simeq \nabla g(\underline{X}_i).$$
The equations then become linear. They have, however, the possibility
of being singular if $\nabla g(\underline{X}_i) = 0$ and, therefore, of being ill-condit-
ioned if this gradient is small.

It is only to be expected that a small gradient of the function g
should lead to problems, since this reflects a lack of distinctive
features in that part of the environment. These particular problems
are fortunately somewhat mitigated when one allows for the errors in
f and g; if, for example, one seeks to minimize the compound expression
$$\int_0^T \underline{v}_m(\underline{x},t)^2 \, dt + \lambda \sum_{i=1}^n \{g(\underline{x}(t_i)) - f(t_i)\}^2 = I_\lambda(\underline{x})$$
subject only to $\underline{x}(0) = \underline{a}$, λ being a positive constant.

The solution to this problem again has $\underline{x}'(t) - \underline{v}(t)$ piecewise
constant; if one makes this assumption and writes $\underline{x}(t_i) = \underline{x}_i$ ($t_0 = 0$),
$t_i - t_{i-1} = h_i$ and $\int_{t_{i-1}}^{t_i} \underline{v}(s) \, ds = \underline{d}_i$, one obtains the nonlinear
algebraic optimization problem of minimizing
$$I_\lambda(\underline{x}) = \sum_{i=1}^n [\{\underline{x}_i - \underline{x}_{i-1} - \underline{d}_i\}^2/h_i + \lambda\{g(\underline{x}_i) - f(t_i)\}^2]$$
with $\underline{x}_0 = \underline{a}$. When one replaces $g(\underline{x}_i)$ by a linear approximation as
before, $I_\lambda(\underline{x})$ becomes a quadratic form in $\underline{x}$ which is positive definite
however small the gradient may be. This leads to a much better condit-
ioned system of linear equations.

The above analysis does no more than scratch the surface of the
problem, which has many more facets, both mathematical and technical.
Mathematically, one may question the assumption we have made concerning
the distribution of $\underline{v}_m$, and particularly its lack of correlation; this
affects the expression to be minimized. Technically there are many
more problems, arising particularly from the fact that the ultimate
object is to design an algorithm to operate in real time with limited
resources. The scheme outlined above requires the solution of n
simultaneous equations, with n increasing with every fresh observation
f of g; one would like a scheme for discarding or merging earlier

observations so that n remained bounded. There is always the chance,
however, that the best fit to the first n+1 observations may differ
markedly from the best fit to the first n - as when one unexpectedly en-
counters, or fails to encounter, a sharp change in the value of g. It
may be that a heuristic approach is the only one possible, but we intend
to pursue mathematically defensible courses as far as we can.

D. C. Handscomb
Oxford University Computing Laboratory
19 Parks Road
Oxford OX1 3PL
England

ISNM, Vol. 67
Numerical Methods of
Approximation Theory, Vol. 7
© 1983 Birkhäuser Verlag Basel

INTERPOLATION AND INSTANT APPROXIMATION

Werner Haußmann

Eberhard Luik

Karl Zeller

Interpolation methods on Chebyshev nodes give rather good
approximations resp. error bounds. We show how certain
improvements can be obtained with little effort: Main
principle FFE (Few Function Evaluations). The remainder
estimate based on one degree of approximation will be refined by introducing several such degrees (in combination
with estimates for Chebyshev coefficients). Next we investigate the question whether the interpolation polynomial
can be abridged (modified to a polynomial of lower degree).
An error bound is obtained by considering the original
nodes and the new coefficients. On the other hand we refine
the interpolation process by introducing two auxiliary
nodes. Finally, we mention modifications, linear programming methods and the multivariate case.

1. Preliminaries

We consider functions f in C[-1,1], the set of real-valued
and continuous functions on the interval [-1,1], endowed
with the norm

$$\| f \| \quad := \quad \sup\nolimits_{|t| \le 1} \; |f(t)|,$$

further polynomials p in $\mathbb{P}_n$ (degree at most n), and the
corresponding degrees of approximation (p_n^* proximum):

$$E_n(f) \quad = \quad \| f - p_n^* \|.$$

To each f we associate the Chebyshev expansion

$$f \;=\; \Sigma' \; a_k T_k \qquad\qquad (\Sigma' \; : \; \text{take } \tfrac{1}{2} a_o T_o),$$

where the coefficients are given by

$$a_k \;=\; \frac{2}{\pi} \int_o^\pi f(\cos t) \, \cos kt \, dt.$$

For convenience, we assume absolute convergence of the
series.

It is a natural task to find good approximations without
much effort or complications; cf. Gutknecht-Trefethen [3],
Haußmann-Luik-Zeller [4,6], Lewanowicz [7], Phillips-
Taylor [8], Scherer-Zeller [10], Watson [11], and the
references given there. We stress here the following
points: FFE (Few Function Evaluations), FFT (Fast Fourier
Transform). We start with a general remainder theorem which
gives information about the accuracy of the approximation
in terms of degrees of approximation. Then we introduce
Chebyshev interpolation (at the zeros of T_{n+1}, later at the
extrema of T_n) and consider the remainder. Next we modify
(abridge) the interpolation polynomial; the new error is
checked at the originally used nodes only. An easily computed bound for the second derivative yields an error
estimate on the full interval. On the other hand we introduce two additional nodes for checking and improving the
accuracy. The final remarks mention certain modifications,
linear programming methods, and the multivariate case.

2. Inequalities

We state inequalities connecting Chebyshev coefficients with norms of functions resp. degrees of approximation. These are useful for general remainder estimates in Section 3 (and for other purposes). First we mention two inequalities based on (discrete) orthogonality:

$$|a_k| \;\le\; \frac{4}{\pi}\, \|f\| \qquad\qquad (k = 1,2,\ldots);$$

$$|a_k| \;\le\; \|p\| \qquad\qquad (p \in \mathbb{P}_{3k-1}; \; k = 1,2,\ldots).$$

If we replace f by $f-p^*_{k-1}$, where p^*_{k-1} is the proximum, then we get

$$|a_k| \;\le\; \frac{4}{\pi}\, E_{k-1}(f) \qquad\qquad (k = 1,2,\ldots).$$

The next estimate comes from Bessel's inequality:

$$\Sigma' \, a_k^2 \;\le\; 2\,\|f\|^2.$$

Using the Cauchy-Schwarz inequality (with factors 1) we get

$$\Sigma'_{(m)} \, |a_k| \;\le\; \sqrt{m}\,(\Sigma' \, a_k^2)^{1/2} \;\le\; \sqrt{2m}\,\|f\|,$$

where m is the number of terms. Sharper inequalities can be obtained by considering (e.g.) $a_k \pm a_{k+1}$:

$$|a_k| + |a_{k+1}| \;\le\; \frac{5}{3}\,\|f\| \qquad (k = 1,2,\ldots);$$

here we have to estimate the integral representation for $a_k \pm a_{k+1}$ with the kernel function $\cos kt \pm \cos(k+1)t$. This method can be refined and extended. In the other direction we have

$$\|f\| \;\le\; \Sigma' \, |a_k|.$$

Finally we mention

$$|a_{k+1}| \;\le\; E_k(f) - (1 + \frac{4}{\pi})\, E_{k+1}(f) \qquad (k = 0,1,\ldots).$$

This is obtained by introducing the proximum p^*_{k+1} and truncating it such that it yields an approximation p_k (thereby using the coefficient inequalities above).

3. Remainder estimates

Suppose R is a continuous linear operator mapping $C[-1,1]$ into itself, having the typical property (for a certain n):

$$R(p) \;=\; 0 \qquad\qquad \text{for } p \in \mathbb{P}_n .$$

In the applications we shall have $R = I - S$, where S is an interpolation operator, I the identity mapping. Clearly,

$$\|R\| \;\leq\; 1 + \|S\| .$$

Introducing the proximum p_n^* we recognize

$$\|R(f)\| \;\leq\; \|R\|\, E_n(f) .$$

This can be refined by using the proximum p_m^* (where $m > n$):

$$p_m^* \;=\; \sum_{k=o}^{m} c_k T_k ; \qquad \|f - p_m^*\| \;=\; E_m(f) .$$

PROPOSITION 1. Given an operator R with the properties as above, and given m with $0 < n < m < 3n$, then the inequality

$$\|R(f)\| \;\leq\; \|R\|\, E_m(f) + \| \sum_{k=n+1}^{m} R(c_k T_k) \|$$

holds with certain coefficients c_k satisfying

$$|c_k| \;\leq\; E_{k-1}(f) + E_m(f) ,$$

$$|c_k - a_k| \;\leq\; \frac{4}{\pi} E_m(f) .$$

Consequently we have

$$\|R(f)\| \;\leq\; \|R\|\, E_m(f) + \sum_{k=n+1}^{m} \|R(T_k)\| \, (E_{k-1}(f) + E_m(f)) .$$

Proof. For the first inequality we write $f = f - p_m^* + p_m^*$. The second one follows from

$$|c_k| \;\leq\; \|p_m^* - p_{k-1}^*\| \;\leq\; \|p_m^* - f\| + \|f - p_{k-1}^*\|$$

using $p_m^* - p_{k-1}^* \in \mathbb{P}_{3k-1}$ and Section 2, which also yields the third inequality. $\square$

In the case of Chebyshev interpolation used below (zeros of T_{n+1}) we have

$$\|R(T_{n+1})\| \;=\; 1, \quad \|R(T_k)\| \;=\; 2 \qquad (n+1 < k < 3n+3) .$$

Thus termwise estimates for the sum above are available; but often the sum can be treated more efficiently. Cf. [5].

4. Chebyshev interpolation

As nodes we use the zeros

$$x_k^{(n+1)} = x_k = \cos t_k, \quad \text{where } t_k = \frac{2k+1}{n+1}\frac{\pi}{2} \quad (0 \le k \le n),$$

of the Chebyshev polynomials T_{n+1}. The interpolation polynomial is given by

$$q_n = S_n f = \sum_{r=0}^{n}{}' b_r T_r$$

with

$$b_r = \frac{2}{n+1} \sum_{k=0}^{n} f(x_k) T_r(x_k) = \frac{2}{n+1} \sum_{k=0}^{n} f(\cos t_k) \cos r t_k$$

(discrete Fourier transform). One has for $r = 0,1,\ldots,n$

$$b_r = a_r - (a_{2n+2-r} + a_{2n+2+r}) + (a_{4n+4-r} + a_{4n+4+r}) - + \ldots,$$

and

$$\|S_n\| \le \frac{2}{\pi} \log(n+1) + 1 \qquad\qquad (n = 0,1,\ldots)$$

(see Fox-Parker [2], p. 67, Rivlin [9], p. 18, and Ehlich-Zeller [1]).

For elucidation we consider $f \in \mathbb{P}_{2n+1}$ (after preliminary approximation). Then we have the error expression

$$f - q_n = a_{n+1}T_{n+1} + \sum_{k=1}^{n} a_{n+1+k}(T_{n+1+k} + T_{n+1-k}),$$

hence (with $x = \cos t$)

$$f(x) - q_n(x) = a_{n+1}\cos(n+1)t$$
$$+ 2\cos(n+1)t \cdot \sum_{k=1}^{n} a_{n+1+k}\cos kt.$$

Our goal is to modify q_n by convenient procedures such that a more favourable error term

$$a_{n+1}T_{n+1} + \sum_{k=1}^{n} (a_{n+1+k}T_{n+1+k} - c_{n+1+k}T_{n+1-k})$$

appears. A good choice is

$$c_{n+1+k} = a_{n+1+k},$$

which is realized in the case of ripple interpolation mentioned in Section 7. But better alternatives are possible; see below and Haußmann-Luik-Zeller [6].

5. Abridged Polynomials

We consider $F \in C[-1,1]$ and $f \in \mathbb{P}_n$ with $\| F-f \| \leq \varepsilon$, where f may be obtained from F by Chebyshev interpolation at the zeros of T_{n+1}. f will be replaced by a $p \in \mathbb{P}_m$ $(m < n)$:

$$p = \sum_{k=0}^{n} c_k T_k \qquad\qquad (c_k = 0 \text{ for } m < k \leq n)$$

(using an approximation process). Then we examine

$$d(t) := f(\cos t) - p(\cos t),$$

where

$$|d''(t)| \leq M := \sum_{k=0}^{n} k^2 |a_k - c_k|.$$

We employ the linear interpolators L_k $(k = -1,0,\ldots,n)$:

$$L_k(t_j) = d(t_j) \qquad\qquad \text{for } j = k, \ k+1,$$

and

$$Q_k(t) := \frac{M}{2}(t-t_k)(t_{k+1}-t) \qquad \text{with } t_k = \frac{2k+1}{n+1}\frac{\pi}{2}.$$

PROPOSITION 2. For $f \in \mathbb{P}_n$, $-1 \leq k \leq n$, and $t_k \leq t \leq t_{k+1}$, the following inequalities hold:

$$|d(t) - L_k(t)| \leq Q_k(t),$$

$$|d(t)| \leq \max_{j=k,k+1} |d(t_j)| + \frac{M}{2}h^2 \qquad (h := \frac{\pi}{2n+2}),$$

$$|d(t)| \leq \max\{|d(t_k)|, |d(t_{k+1})|, |L_k(\frac{t_k+t_{k+1}}{2})| + Mh^2\}.$$

Proof. The first inequality derives from the usual interpolation remainder formula. In addition, we observe that

$$|L_k(t)| \leq \max_{j=k,k+1} |L_k(t_j)| = \max_{j=k,k+1} |d(t_j)|$$

and

$$|Q_k(t)| \leq \frac{M}{2}\frac{(t_{k+1}-t_k)^2}{4} = \frac{M}{2}h^2,$$

which yields the second inequality. In order to get the third one we replace Q_k by the majorant given by its tangents at the endpoints t_k and t_{k+1}. $\square$

The exact maximum of $L_k + Q_k$ (if located in the interior) is given by

$$[L_k + Q_k](\frac{t_k + t_{k+1}}{2}) + \frac{1}{2M}(L_k')^2 \qquad\qquad (M \neq 0).$$

6. Refined interpolation

We interpolate $f \in \mathbb{P}_{2n+1}$ at the zeros of T_{n+1} obtaining q_n. Then we evaluate the error $d := f - q_n$ at two extra points, namely $x = 1$ ($t = 0$) and $x = -1$ ($t = \pi$). This yields

$$d(1) = a_{n+1} + 2a_{n+2} + 2a_{n+3} + \cdots ,$$

$$(-1)^{n+1}d(-1) = a_{n+1} - 2a_{n+2} + 2a_{n+3} - + \cdots .$$

These equations give us an approximate information about the two leading coefficients. In many cases we can improve the accuracy by changing the leading coefficient in q_n. Let us point out this for an $f \in \mathbb{P}_{n+2}$. Then we can compute a_{n+1} and a_{n+2} exactly. Put

$$b_n^\# = b_n + 2a_{n+2} = a_n + a_{n+2},$$

and, more general,

$$b_n^* = a_n + Aa_{n+2},$$

where

$$A^{-1} := 1 + 2|a_{n+2}/a_{n+1}| \qquad \qquad (\text{if } a_{n+1} \neq 0).$$

Replacing the leading coefficient b_n in q_n by one of these values we get $q_n^\#$ resp. q_n^*.

PROPOSITION 3. For $f \in \mathbb{P}_{n+2}$ the polynomials $q_n^\#$ resp. q_n^* obtained in this way satisfy the error estimates

$$\| f - q_n^\# \|^2 \leq a_{n+1}^2 + 4a_{n+2}^2 ,$$

$$\| f - q_n^* \|^2 \leq a_{n+1}^2 + 5Aa_{n+2}^2 + 4A^2 a_{n+2}^4 / a_{n+1}^2 .$$

Proof. We have (with $x = \cos t$)

$$(f - q_n^\#)(x) = a_{n+1}\cos(n+1)t - 2a_{n+2}\sin(n+1)t \sin t.$$

This yields the first estimate. The polynomial q_n^* can be considered as a weighted average of the two polynomials using $b_n^\#$ and $b_n' = a_n - a_{n+2}$. The weights are chosen in a way such that the complementary behaviour of the two error curves is utilized. Further details are given in [6]. $\square$

The attenuation factor A is particularly useful if the value $|a_{n+2}/a_{n+1}|$ is around one half.

7. Remarks

In a similar way we can treat the interpolation based on the extrema of T_n. Essentially one can replace the main error coefficient $2a_{n+1}$ by a_{n+1}, or achieve a result as in Proposition 3. But it might be better to use the extrema of T_{n+1} and then omit the term $b_{n+1}T_{n+1}$. The latter one is then the main error term (ripple interpolation). The other b_k yield nice approximation properties (cf. Section 4). Further one could use more than two extra nodes; but methods as in Section 5 seem to be preferable. Finally one might consider nodes giving smaller operator norms (but these are apparently more difficult to handle).

A suitable polynomial p in Section 5 can be obtained in several ways: approximation operators (possibly nonlinear as in Section 6) or exchange resp. optimization algorithms (with a fixed grid containing few points: $x_o, \ldots, x_n$). For instance one could (nearly) minimize a variable u under the constraints

$$u \geq |d(x_k)| ; \quad \frac{1}{2}|d(x_k) + d(x_{k+1})| + Mh^2 + \theta.$$

Here Mh^2 comes from Proposition 2; and θ accounts for possible changes in M during the process. This can be refined (cf. Section 5). Further one could use typical correcting polynomials which decrease one hump (perhaps more humps) of the error function (cf. Scherer-Zeller [10]); but the corresponding theory has yet to be developed.

Our methods also apply to multivariate approximation, primarily in the tensor product case; modifications (variable transformation, imbedding domains, use of other basic functions etc.) lead to more general considerations. For these cases the FFE (Few Function Evaluations) principle is especially important. The search for a maximum in a multidimensional domain might be very time-consuming. To obtain a proximum (or a corresponding H-set) is often a complicated task; then one could be content with a "good" approximation rather than with a best one.

References

[1] Ehlich, H., Zeller, K.: Auswertung der Normen von Interpolationsoperatoren. Math. Annalen 164 (1966), 105-112.

[2] Fox, L., Parker, I. B.: Chebyshev Polynomials in Numerical Analysis. London, Oxford University Press 1968.

[3] Gutknecht, M. H., Trefethen, L. N.: Real Polynomial Chebyshev Approximation by the Carathéodory-Fejér Method. SIAM J. Numer. Anal. 19 (1982), 358-371.

[4] Haußmann, W., Luik, E., Zeller, K.: Biorthogonality in Approximation. Internat. Ser. Numer. Math. Vol. 61 (1982), 185-189.

[5] Haußmann, W., Luik, E., Zeller, K.: Cubature Remainder and Biorthogonal Systems. Internat. Ser. Numer. Math. Vol. 61 (1982), 191-200.

[6] Haußmann, W., Luik, E., Zeller, K.: Bequeme Verfahren zur Approximation. Z. Angew. Math. Mech. 63 (1983), T349-T350.

[7] Lewanowicz, S.: Some Polynomial Projections with Finite Carrier. J. Approx. Theory 34 (1982), 249-263.

[8] Phillips, G. M., Taylor, P. J.: Polynomial Approximation Using Equioscillation on the Extreme Points of Chebyshev Polynomials. J. Approx. Theory 36 (1982), 257-264.

[9] Rivlin, T. J.: The Chebyshev Polynomials. New York, John Wiley & Sons 1974.

[10] Scherer, R., Zeller, K.: Floppy vs. Fussy Approximation. Internat. Ser. Numer. Math. Vol 59 (1982), 171-178.

[11] Watson, G. A.: Approximation Theory and Numerical Methods. Chichester, John Wiley & Sons 1980.

Werner Haußmann
Department of Mathematics
University of Duisburg
D-4100 Duisburg

Eberhard Luik
Karl Zeller
Department of Mathematics
University of Tübingen
D-7400 Tübingen

ISNM. Vol. 67
Numerical Methods of
Approximation Theory, Vol. 7
© 1983 Birkhäuser Verlag Basel

ON THE USE OF CERTAIN PROPERTIES
OF STIFFNESS MATRICES IN FINITE ELEMENT METHODS

H. Kardestuncer

The stiffness matrices encountered in the analysis of elastic systems using finite element methods possess certain properties the use of which in the solution procedures reduces the computer time and increases the accuracy of the results. The presentation places emphasis on the application of the theory to skeletal systems using one dimensional elements. Reference, however, has been made to two and three dimensional finite elements.

Introduction

The complexities of problems in science and engineering today force most researchers to emphasize approximate methods. Although approximate methods have been known to mankind much earlier than analytical ones, their popularity had to wait for the advancement of the necessary tools (the computer, in this case) which only occurred during the past two to three decades. The primary reason for this is that most approximate methods often involve solution of a set of simultaneous equations. The order of such equations for problems in practice is very large. For some problems, even the biggest computer today is not big enough. Regardless of the nature of the problem and the method employed for its solution, in engineering we often encounter the following equation.

$$P = K U \tag{1}$$

If the problem is in solid mechanics and the technique is the finite element method (FEM), the entities in this equation represent force, stiffness, and displacement matrices, respectively. For the majority of problems, matrix K

is sparse, symmetric, and banded. Although these properties of K facilitate
the solution procedure somewhat, very often the order of such a matrix is
quite large. For example, a three-dimensional finite element model of a cube
with only 20 nodes along each edge and 3 degrees of freedom per node results
in about 24,000 equations. This in turn requires about 29 million words of
storage which far exceeds the core limitations of any machine in existence[1].
Another example cited in [1] which deals with the 3-D turbulent unsteady flow
using the finite difference method (FDM) with 128^3 mesh points with 4 degrees
of freedom per node (velocity and pressure) requires 11 trillion words of storage. In the latter case, the matrix is also non-symmetric.

It is certainly known that, in contrast to most other approximate
methods, e.g. Rayleigh-Ritz, Galerkin, Kantorovich-Krylov, or even FDM, the
finite element method requires much larger numbers of equations to solve. The
applicability of those methods, however, is not as versatile as the finite element methods. For this reason, in this paper, we shall only consider those
entities that are encountered in FEM in solid mechanics.

There are primarily two distinct areas where the results of FEM
can be improved: first, the improvements of the approximate functions at the
element domain; and second, the improvements during the solution procedures
at the global domain. We shall not deal here with the numerous ways of attaining these improvements; instead, we shall point out certain properties of
entities and try to utilize them to improve or facilitate the results.

Element Domain

One of the nice features of FEM is the removal of the conditions
posed on the approximate function to satisfy the governing differential equation and on the prescribed boundary conditions a priori. Once the local function is chosen, the physical phenomenon is normally expressed as in (2).

$$P_e = k_e u_e \tag{2}$$

This is similar to Eq. (1) yet it is confined to the element domain. Since
Eq. (2) is in matriciel form, it requires the identification of local coordinate axes prior to choosing the approximate function. It is at this point,
we believe, that very little attention is paid to the physical nature of the
problem. For instance, for a line element, the possible choices of local coordinate axes are shown in Fig. (1).

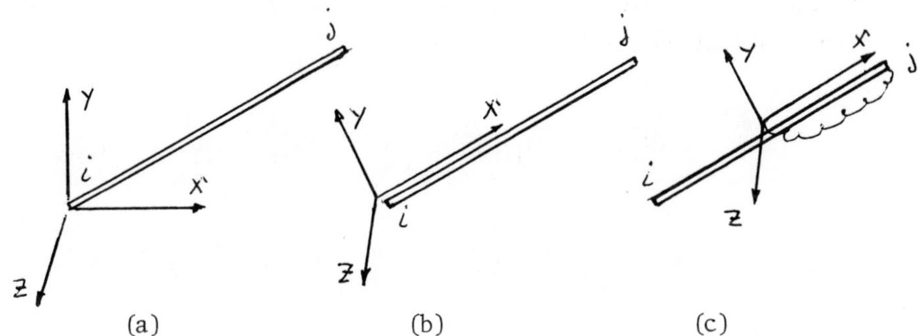

(a)   (b)   (c)

Fig. (1) Local Coordinate Axes for a Line Element

Of these three different choices (a) is the least suited, (b) is the most common, and (c) is the best choice. If one uses a cubic approximating function

$$\tilde{u}_e = N(x)\, U_e \tag{3}$$

where

$$N(x) = c_i\, x^i \quad , \quad i = 0, \ldots, 3$$

then, depending upon the choices shown in Fig. (1), the resulting stiffness matrix k_e would be full, semi-full, or diagonal, respectively. This is due to the fact that the entities in Eq. (2) are second and fourth order tensors instead of ordinary vectors and matrices [2]. The coordinate system shown in Fig. (1c), in fact, represents the principle coordinate axes of the stiffness tensor. The tensorial properties of this entity in Eq. (2) do not come to the surface if one uses the approximation shown in Eq. (3) without making proper reference to the coordinate axes and transfer functions. Instead, at least for solid mechanics problems, we shall refer to Castigliano's theorems or, more generally, to Menabrea's least energy principle, from the beginning of the formulation and we shall observe the tensorial properties of each entity. Let, for instance, the total potential energy* of the element be written as

$$\pi = \frac{1}{2} p^{iq} u_{iq} - p^{mq} u_{mq} - p^{sq} u_{sq} \qquad (4)$$

While the first term in this equation represents the strain energy of the system, the second and third terms represent the body and surface force potentials. The minimization of this equation yields

$$\frac{\delta \pi}{\delta u_{jr}} = \frac{1}{2} \frac{\delta p^{iq}}{\delta u_{jr}} u_{iq} + \frac{1}{2} p^{iq} \frac{\delta u_{iq}}{\delta u_{jr}} - \frac{\delta p^{mq}}{\delta u_{jr}} u_{mq}$$

$$- p^{mq} \frac{\delta u_{mq}}{\delta u_{jr}} - \frac{\delta p^{sq}}{\delta u_{jr}} u_{sq} - p^{sq} \frac{\delta u_{sq}}{\delta u_{jr}} = 0$$

Since

$$\frac{\delta u_{iq}}{\delta u_{jr}} = 1 \qquad \text{for } i,q = j,r ; \quad \text{zero otherwise}$$

the above equation reduces to

$$\frac{1}{2} p^{iq} + \frac{1}{2} k^{iqjr} u_{iq} - k^{mqjr} u_{mq} - k^{sqjr} u_{sq} = 0$$

where

$$k^{kqjr} = \frac{\delta p^{iq}}{\delta u_{jr}}$$

with similar substitutions for other terms. Considering, however, that m+s=i then

$$p^{iq} = k^{iqjr} u_{jr}$$

$$\qquad (5)$$

This is the basic equation established in the element domain. It is similar to the following well-known tensorial equation in solid mechanics.

---

* We admit that for many problems in practice, one can not establish a suitable energy functional.

$$\sigma^{iq} = E^{iqjr}\,\varepsilon_{jr} \tag{6}$$

While Eq. (6) is purely physical, Eq. (5) is geometrical as well as physical. Both equations obey the tensorial transformation rules. In the former, however, the transformations are more general (rotational as well as translational). Because of this, the determination of principle directions of the stiffness matrix (bivalent version of the stiffness tensor) involves translational transformations. Such transformations can be written as

$$k^{i'q'j'r'} = \frac{\delta x^{i'}}{\delta x^{i}}\,\frac{\delta x^{q'}}{\delta x^{q}}\,\frac{\delta x^{j'}}{\delta x^{j}}\,\frac{\delta x^{r'}}{\delta x^{r}}\,k^{iqjr} \tag{7}$$

In matrix notation, this will take the following form

$$k' = H^* \, R^* \, k \, R \, H \tag{8}$$

in which H and R matrices represent translational and rotational transformations respectively.

Global Domain

Once the stiffness matrix is composed in reference to the principle coordinate axes for the elements, the next step is the assembly of the system by satisfying certain conditions at the interfaces (compatibility and equilibrium in the case of solid mechanics problems) between the elements. For the simplest case where the line elements are connected in series, the stiffness tensor of single elements can be translated from the element principle coordinates to the principle coordinates for the entire system. The assembled stiffness matrix remains diagonal and the solution of the system requires inversion of matrices the order of which is the same as that of element matrices. The assembly of matrices for series members requires translational transformation and summation of the element flexibility matrices.

$$K_{eq} = [\, \Sigma \, H^* \, D \, H \,]^{-1} \tag{9}$$

The flexibility matrices D (the inverse of the stiffness matrices) for the elements are often known explicitly.

Equation (9) replaces all elements in series with an equivalent matrix of the same order. Similar replacement can be done for elements in parallel connections. Thus

$$K_{eq} = \Sigma\, K_e \tag{10}$$

in which case the equivalent stiffness matrix is equal to the sum of the stiffnesses of each element.

Once the equivalent stiffness matrices for the elements in series and parallel are established, the combination of Eqs. (9) and (10) will produce the equivalent matrices for closed-loops[3]. As a result of this, the order of the overall stiffness matrix (Eq. 1) for the entire system will be considerably reduced. Numerical examples supporting this claim can be found in [4].

The concept presented here for line elements in solid mechanics is also applicable, with certain care, to higher order elements. In this case, for example, the following integration of the approximate function for the elements should be done in the principle axes of the elements. Thus

$$K^{iqjr} = \int_V b_r^q\, D^{iqjr}\, b_q^r\, dv \tag{11}$$

in which D^{iqjr} is the material tensor (not to be confused with the flexibility tensor in Eq. (9)) and

$$b_r^q = \frac{1}{2} \left(\frac{\delta}{\delta x_q} + \frac{\delta}{\delta x_r} \right) N\, (x_q,\, x_r) \tag{12}$$

where $N(X_q,\, X_r)$ is the shape function. In the case of elements with curved boundaries, one may attempt to choose the local coordinate axes such that, in reference to them the geometry of the element becomes linear. Depending upon the desired continuity requirements at the interfaces, the shape function can also be chosen as identical with the element geometry. Such an element is often referred to as an isoparametric element. If, on the other hand, the shape functions are chosen to be of higher or lower order than the geometry of the element, such elements are called super-parametric and sub-parametric respectively. In all cases, however, once the integration in Eq. (11) is done, the transformation of the stiffness tensor from the chosen curvi-linear coordinates to the global coordinates (most often rectangular) becomes necessary. Such a transformation, which follows one of the standard tensor transformations, is always valid regardless of the chosen local or global coordinate systems.

Needless to say, the question of the most suitable coordinate system for a given element (in particular, for the curved elements) will always remain to be determined. A thorough knowledge of the physical problem and of tensor transformations is very valuable in finding the answers to this question.

<u>Numerical Example</u>

Consider a small problem which consists of four line elements as shown in Fig. (2).

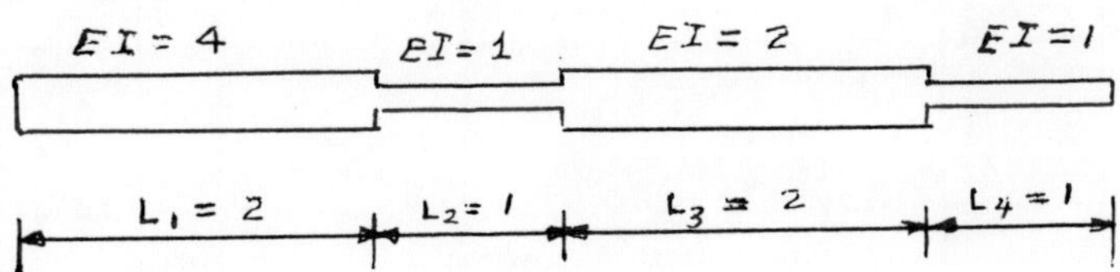

Fig. (2)   One-dimensional elements in series

The governing differential equation and the boundary conditions are

$$\frac{d^4u}{dx^4} = \frac{p}{EI}$$

$$u(0) = \dot{u}(0) = \ddot{u}(L) = 0 \ , \ \dddot{u}(L) = 10$$

If one uses the cubic approximation given in Eq. (3) for the elements, the final set of equations for the entire system after the introduction of the boundary conditions will be

$$
\begin{bmatrix} 0 \\ 0 \\ 0 \\ 0 \\ 0 \\ 0 \\ 0 \\ 10 \end{bmatrix}
=
\begin{bmatrix}
18 & 0 & -12 & -6 & 0 & 0 & 0 & 0 \\
0 & 12 & 6 & 2 & 0 & 0 & 0 & 0 \\
-12 & 6 & 15 & 3 & -3 & 3 & 0 & 0 \\
-6 & 2 & 3 & 8 & 3 & 2 & 0 & 0 \\
0 & 0 & -3 & 3 & 15 & -3 & -12 & -6 \\
0 & 0 & -3 & 2 & -3 & 8 & 6 & 2 \\
0 & 0 & 0 & 0 & -12 & 6 & 12 & 6 \\
0 & 0 & 0 & 0 & -6 & 2 & 6 & 4
\end{bmatrix}
\begin{bmatrix} u_2 \\ \dot{u}_2 \\ u_3 \\ \dot{u}_3 \\ u_4 \\ \dot{u}_4 \\ u_5 \\ \dot{u}_5 \end{bmatrix}
$$

This equation can, of course,, be solved by any existing method. Instead, however, we shall employ Eq. (8) to each element prior to assembling them which in turn will result in the following set instead of in the above equation. Notice that the new set is already in block diagonal form.

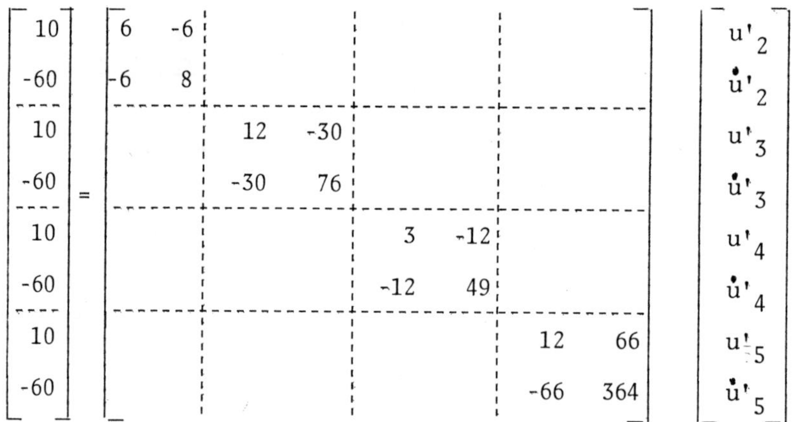

In this case, H and R matrices are

$$H_i = \begin{bmatrix} 1 & 0 \\ L_i & 1 \end{bmatrix} \qquad\qquad R = I$$

Notice that in this specific problem, the final set does not necessitate the use of any solution procedures because the principle sub-matrices are already uncoupled. The resulting vector U' and its transformed version U which is the solution to the original set are shown below.

$$U' = \begin{bmatrix} -23.3 \\ -25 \\ -86.5 \\ -35 \\ -76.5 \\ -20 \\ -26.7 \\ -5 \end{bmatrix} \qquad \therefore \qquad U = H^* U' = \begin{bmatrix} 26.6 \\ -25 \\ 70 \\ -60 \\ 215 \\ -80 \\ 297 \\ -85 \end{bmatrix}$$

Further details of this example can be found in Reference [5]. We must at this point state very clearly the fact that, for most problems, the solution vector U can not always be obtained by predetermined H operations. H operations reduce the band width by one (in terms of block matrices). If the original set is a three-diagonal block matrix, as in this example, then the solution vector is the outcome of the H operations. Very few problems in practice fall into this category. The reduction of band width by means of H operations, however, is only one step toward the final solution. This is because the finite element analysis of elastic systems often results in equicofactor stiffness matrices[6].

Conclusion

In this presentation, we have pointed out briefly certain properties of entities encountered in the finite element methods for the purpose of attaining the results more efficiently and easily. Although there exist many methods and algorithms for the solution of a set of simultaneous equations, none, in the opinion of the author, pays attention to the physical nature of the problem. Since all of these methods, whether direct or iterative, contain only slight modifications of Gauss's method made to accomodate sparseness, symmetry, band, three-diagonal, etc., the author believes that not much progress has been made since Gauss. Much emphasis, on the part of engineers, has been laid on the formulation of the problems. Engineers have been content to employ methodologies which are equally suited to physical or mathematical problems. Mathematicians, on the other hand, pay little attention to the physical nature of the problem at this stage. In recent years, because of the popularity of approximate methods in the physical sciences, however, some dialogue between mathematicians and engineers has begun. Such a dialogue has not yet reached the solution procedure and much of the accomplishment in this respect has been mechanical rather than mathematical. Not the methods but the tools have registered improvements (e.g. speed, capacity, and the word length of computers). Engineers and mathematicians have taken these improvements for granted and have been able to handle larger and larger problems by making finer approximations in respect to local approximate functions or to mesh size. We have now reached the point where improvements in tools do not improve the results as much as they did in the past. We must now develop problem-oriented methodologies. The solution procedures of two sets of simultaneous equations, one resulting from solid mechanics, the other from fluid mechanics, for example, need not be identical. One way to accomplish this is to pay attention to the tensorial and physical properties of entities in each problem and make use of tensorial invariants not only during the formulations but during the solutions as well. The author hopes that this presentation is a very modest attempt in this direction.

References

1. Hughes, Thomas J.R., Itzhak Levit, and James Winger: An Element-by-Element Solution Algorithm for Problems of Structural and Solid Mechanics. Computer Methods in Applied Mechanics and Engineering 36 (1983), 241-254.

2. Kardestuncer, H.: Tensors versus Matrices in Discrete Mechanics. In: Problem Analysis in Science and Engineering, F. H. Branin, Jr. and K. Huseyin (eds.). New York: Academic Press 1977.

3. Kardestuncer, H.: Equivalent Stiffness-Flexibility Matrices of Closed-Loops Subject to Temperature Variation. Proceedings of VI IKM, Weimar, 1972.

4. Kardestuncer, H.: Elementary Matrix Analysis of Structures. New York: McGraw-Hill Book Company 1974.

5. Kardestuncer, H.: Finite Elements via Tensors. Vienna and New York: Springer Verlag CISM Series 1092.

6. Kardestuncer, H.: On the Generalization of Equicofactor Matrices. International Journal for Numerical Methods in Engineering 7 (1973), 491-496.

H. Kardestuncer
University of Connecticut
U-37
Storrs, CT 06268
USA

ISNM. Vol. 67
Numerical Methods of
Approximation Theory, Vol. 7
© 1983 Birkhäuser Verlag Basel

EULER-FROBENIUS-POLYNOME

Walter Schempp

Based on the theory of cardinal exponential splines this paper presents a useful complex contour integral representation with non-compact integration path for the Euler-Frobenius polynomials. These polynomials are well known from the theory of attenuation factors in numerical Fourier analysis. It is shown that the contour integral approach to the Euler-Frobenius polynomials allows to deduce in a simple way all their fundamental properties.

Die nach Euler und Frobenius benannten Polynome $(p_m)_{m \geq 1}$ wurden 1755 von L. Euler eingeführt [1] und in einer grundlegenden Arbeit aus dem Jahre 1910 von F.G. Frobenius im Zusammenhang mit den Bernoullischen Zahlen näher untersucht [2]. Frobenius stellte auch fest, daß die Polynome $(p_m)_{m \geq 1}$ der nicht-linearen Funktionalgleichung

$$h^{m-1} p_m \left(\frac{1}{h}\right) - p_m (h) \qquad (h \neq 0)$$

genügen ("Reflexivitätseigenschaft"; vgl. Korollar 1 zum unten stehenden Satz). Später gewannen die Euler-Frobenius-Polynome $(p_m)_{m \geq 1}$ neue Bedeutung für Probleme der kardinalen Spline-Interpolation (Schoenberg [6]) und der numerischen Fourier-Analysis (Theorie der Abminderungsfaktoren; vgl. Quade-Collatz [3]).

Das Ziel dieses Beitrags ist es, einen neuen Zugang zu den Euler-Frobenius-Polynomen $(p_m)_{m \geq 1}$ aufzuzeigen, der auf der Theorie der <u>kardinalen Exponentialsplines</u> beruht und zu einer komplexen Kurvenintegraldarstellung mit nichtkompaktem Integrationsweg führt (vgl. [4] und die Monographie [5]). Aus ihr lassen sich dann leicht die fundamentalen Eigenschaften der Polynome $(p_m)_{m \geq 1}$ ablesen.

Für jede ganze Zahl $m \geq 1$ bezeichne $\mathsf{G}_m(\mathbb{R}\,;\mathbb{Z})$ den komplexen Vektorraum der kardinalen Spline-Funktionen vom Grad m auf $\mathbb{R}$ zum Knotengitter $\mathbb{Z}$. Die Elemente von $\mathsf{G}_m(\mathbb{R}\,;\mathbb{Z})$ sind also $(m-1)$-mal stetig differenzierbare, komplexwertige Funktionen auf $\mathbb{R}$, deren Einschränkungen auf die Teilintervalle $[n,n+1]$, $n \in \mathbb{Z}$, durch Polynome höchstens vom Grad m mit komplexen Koeffizienten gegeben werden. Eine ausgezeichnete $\mathbb{C}$-Basis von $\mathsf{G}_m(\mathbb{R}\,;\mathbb{Z})$ wird durch die "verschobenen" Funktionen

$$\mathbb{R} \ni x \rightsquigarrow b_m(x-n) \in \mathbb{R}_+ \qquad (n \in \mathbb{Z})$$

geliefert, wobei der kardinale Spline $b_m \in \mathsf{G}_m(\mathbb{R}\,;\mathbb{Z})$ durch die beiden folgenden Eigenschaften charakterisiert wird: Es gilt

$$(i) \qquad \mathrm{Supp}(b_m) = [0,m+1] \qquad (\text{"Trägerbedingung"}),$$

und

$$(ii) \qquad \int_{\mathbb{R}} b_m(t)\,dt = 1 \qquad (\text{"Standardisierungsbedingung"}).$$

Man nennt b_m einen <u>kardinalen Basis-Spline</u> (shift 0) vom Grad m. Mit Hilfe dieses Begriffes kann folgende Definition gegeben werden.

DEFINITION. Für jede natürliche Zahl $m \geq 1$ heißt das durch die Gleichung

$$p_m(h) = m! \sum_{0 \leq n \leq m-1} b_m(n+1)h^n$$

gegebene Polynom $p_m \in \mathbb{Z}[h]$ das m-te Euler-Frobenius-Polynom.

Aus den bekannten Eigenschaften des kardinalen Basis-Splines b_m folgt, daß die Koeffizienten von p_m natürliche Zahlen > 0 sind, daß der Höchstkoeffizient $=1$ ist und ebenso

$$p_m(0) = 1$$

gilt. Der genaue Grad von p_m ist also m-1.

Sei jetzt $h \neq 0$ eine feste komplexe Zahl. Im Vektorraum $\mathsf{G}_m(\mathbb{R};\mathbb{Z})$ bilden diejenigen Spline-Funktionen, welche der inhomogenen linearen Differenzengleichung mit konstanten Koeffizienten

$$f(x+1)-hf(x)=0 \qquad (x \in \mathbb{R})$$

genügen, einen eindimensionalen Untervektorraum. Man nennt diese Splines kardinale Exponentialsplines. Sie haben die Form

$$s_m = C_m \sum_{n \in \mathbb{Z}} h^n b_m(.-n)$$

mit einer Konstanten $C_m \in \mathbb{C}$. Demnach hängt das Euler-Frobenius-Polynom p_m eng mit $s_m(0)$ zusammen. Vom kardinalen Basis-Spline b_m weiß man aber, daß er sich als (endliche) Linearkombination "abgeschnittener" Potenzen (truncated powers) darstellen läßt. Für diese stehen, über den Inversionssatz für die Laplace-Transformation, die Dirichletschen Darstellungsformeln zur Verfügung.

Mit ihnen erhält man den nachstehenden

SATZ.   Sei $h \in \mathbb{C}$ eine komplexe Zahl mit $h \neq 0$, $|h| \neq 1$. Dann besitzt $p_m(h)$ die komplexe Kurvenintegraldarstellung

$$p_m(h) = \frac{(h-1)^{m+1}}{h} \; \frac{m!}{2\pi i} \int_P \frac{e^z}{(e^z-h)\,z^{m+1}} \, dz \qquad (m \geq 1)$$

wobei P den orientierten Rand eines abgeschlossenen, vertikalen Streifens in der offenen rechten bzw. linken komplexen Halbebene bezeichnet, je nachdem ob $|h| > 1$ oder $0 < |h| < 1$ ist, welcher die "kritische" Gerade $\operatorname{Re} z = \log |h|$ in seinem Innern enthält.

Für den Integrationsweg P sind also die nachstehenden Konstellationen möglich:

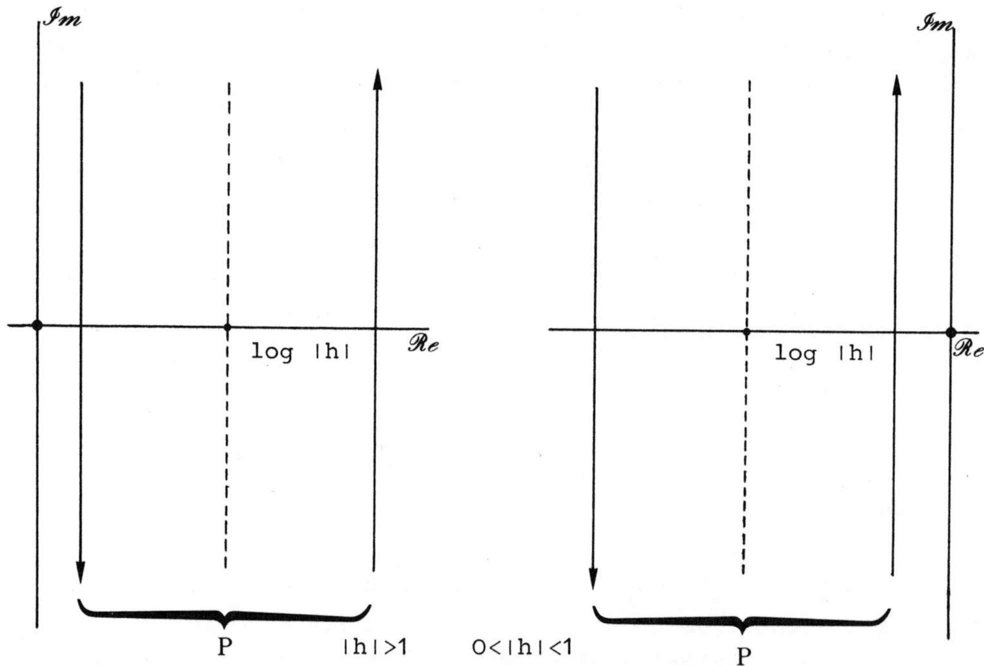

Einen Beweis für den vorstehenden Satz findet man in [4] und [5]. Die Symmetrie zwischen den beiden Fällen $|h|>1$ und $0<|h|<1$ liefert das

KOROLLAR 1 (Frobenius).   <u>Für alle</u> $h \in \mathbb{C}$ <u>mit</u> $h\neq 0$ <u>gilt die Funktionalgleichung</u>

$$h^{m-1} p_m\left(\tfrac{1}{h}\right) = p_m(h) \qquad (m\geq 1),$$

<u>d.h. die Euler-Frobenius-Polynome</u> $(p_m)_{m\geq 1}$ <u>sind reflexiv.</u>

Mit Hilfe der Cauchyschen Integralformel folgt aus obigem Satz ferner das

KOROLLAR 2 (Euler).   <u>Für alle</u> $h \in \mathbb{C}$, $h\neq 0$ <u>und</u> $|h|\neq 1$ <u>gilt</u>

$$\frac{(h-1)e^z}{h(h-e^z)} = \sum_{m\geq 0} \frac{p_m(h)}{(h-1)^m}\,\frac{z^m}{m!} \qquad (\,|z|<|\log |h||\,)$$

<u>mit</u> $p_0(h) = \tfrac{1}{h}$ ("Erzeugende Funktion").

Koeffizientenvergleich liefert

$$(h-1)\sum_{\nu\geq 0}(\nu+1)^m h^{-2-\nu} = \frac{p_m(h)}{(h-1)^m} \qquad (m\geq 0).$$

Wendet man sodann die Funktionalgleichung des Korollars 1 an, so findet man das

__KOROLLAR 3.__ __Für__ __alle__ $h \in \mathbb{C}$ __mit__ $h \neq 1$ __gilt__ __die__ __formale__ __Potenzreihenentwicklung__

$$\frac{p_m(h)}{(1-h)^{m+1}} = \sum_{n \geq 0} (n+1)^m h^n \qquad (m \geq 1).$$

Die vorstehende Potenzreihenentwicklung ergibt das

__KOROLLAR 4__ (Euler). __Die__ __Euler-Frobenius-Polynome__ __genügen__ __der__ __Rekursionsgleichung__

$$p_{m+1}(h) = (mh+1)p_m(h) - h(h-1)p_m'(h) \qquad (m \geq 1).$$

Man erhält insbesondere

$$p_m(1) = m! \qquad (m \geq 1).$$

Mit Hilfe des vorstehenden Korollars lassen sich die Polynome $(p_m)_{m \geq 1}$ leicht explizit berechnen. Setzt man

$$p_m(h) = \sum_{0 \leq n \leq m-1} a_n^{(m)} h^n \in \mathbb{Z}[h] \qquad (m \geq 1),$$

so erfüllen die ganz-zahligen Koeffizienten $a_n^{(m)} = m! b_m(n+1)$ die Beziehungen

$$a_0^{(m)} = 1, \quad a_n^{(m)} = a_{m-n-1}^{(m)}, \quad a_{m-1}^{(m)} = 1$$

und die Dreiecksmatrix $(a_n^{(m)})_{\substack{n \geq 0 \\ m \geq 1}}$ nimmt folgende Gestalt an:

m \ n	0	1	2	3	4	5	6
1	1						
2	1	1					
3	1	4	1				
4	1	11	11	1			
5	1	26	66	26	1		
6	1	57	302	302	57	1	
7	1	120	1191	2416	1191	120	1

Aus der Rekursionsgleichung folgt durch vollständige Induktion nach m das

KOROLLAR 5 (Frobenius). Für jede natürliche Zahl $m \geq 2$ sind die Nullstellen des Euler-Frobenius-Polynoms p_m einfach und liegen auf der offenen negativen reellen Halbgeraden. Mit h_0 ist auch $\frac{1}{h_0}$ eine Nullstelle von p_m und $p_m(-1) = 0$ gilt genau dann, wenn m eine gerade Zahl ist.

Zusammenfassend kann festgestellt werden, daß es die im Satz angegebene komplexe Kurvenintegraldarstellung der Euler-Frobenius-Polynome $(p_m)_{m \geq 1}$ mit nichtkompaktem Integrationsweg P erlaubt, die fundamentalen Eigenschaften der Polynomfamilie $(p_m)_{m \geq 1}$ auf einfache Weise zu gewinnen.

Literatur

1. Euler, L.: Institutiones calculi differentialis cum eius usu
    in analysi finitorum ac doctrina serierum. Academiae Imperi-
    alis Scientiarum Petropolitanae 1755. In: Opera omnia, Series
    I (Opera Mathematica), Vol. X. G. Kowalewski editor.
    Leipzig-Berlin: B.G. Teubner 1913

2. Frobenius, F.G.: Über die Bernoullischen Zahlen und
    Eulerschen Polynome. Sitzungsberichte der Königlich Preußischen
    Akademie der Wissenschaften zu Berlin (1910), 809-847. In:
    Gesammelte Abhandlungen, Vol. III. J-P. Serre, editor.
    Berlin-Heidelberg-New York: Springer 1968

3. Quade, W., Collatz, L.: Zur Interpolationstheorie der reellen
    periodischen Funktionen. Sitzungsberichte der Preuss. Akad.
    der Wiss., Phys.-Math. Kl. XXX (1938), 383-429

4. Schempp, W.: A contour integral representation of Euler-Fro-
    benius polynomials. J. Approx. Theory $\underline{31}$ (1981), 272-278

5. Schempp, W.: Complex Contour integral representation of car-
    dinal spline functions. Contemporary Mathematics, Vol. $\underline{7}$.
    Providence, R.I.: Amer. Math. Soc. 1982

6. Schoenberg, I.J.: Cardinal spline interpolation. Regional
    Conference Series in Applied Mathematics, Vol. $\underline{12}$.
    Philadelphia, PA: Society for Industrial and Applied
    Mathematics 1973

Lehrstuhl für Mathematik I

der Universität Siegen

Hölderlinstraße 3

D-5900 Siegen

ISNM, Vol. 67
Numerical Methods of
Approximation Theory, Vol. 7
© 1983 Birkhäuser Verlag Basel

ZUR DIAGNOSE VERLETZTER KNIE
– EIN IDENTIFIKATIONSPROBLEM –

Helmut Werner

In this paper we report on an identification problem from medicine. We consider
the following problem: Given a certain number of types of injuries of the knee
and some well defined (discrete) set of tests and the associated answers. Find
a minimal subset of the tests such that from their outcome (i.e. answers) each
specific type of injury is uniquely determined.
We describe how to determine the number of tests necessary for the solution
of the problem and also how to find the set of all solutions. The method is
demonstrated for a simple model problem. Additional problems arising in real
life medical applications are only pointed out. They are treated in another
paper to appear.

In diesem Vortrag wird über ein Identifikationsproblem aus dem Bereich
der Medizin berichtet. Für eine Anzahl von Verletzungstypen liegen für
wohldefinierte Tests die zugehörigen diskreten Antworten vor. Es ist eine
Menge von Tests zu bestimmen, aus deren Ergebnissen eindeutig auf den vorliegenden Verletzungstyp geschlossen werden kann. Es wird beschrieben, wie
man die notwendige Anzahl von Tests und die Gesamtheit der Lösungsmengen
mit dieser Anzahl von Tests aufstellen kann. Das Verfahren wird an einem
Modellfall demonstriert, auf die weiteren in der Praxis auftretenden Probleme
wird hier nur hingewiesen .

Medizinische Fragestellung

Vor einigen Monaten wurde ich von Herrn Dr. med. Stedtfeld, Universitäts-Klinik Münster um Unterstützung bei folgender Problemstellung gebeten:
Knie mit verletzten Bändern sind nicht mehr so stabil wie gesunde Knie, sondern zeigen bei Beugung des Knies und Drehung des Fußes unter einer bei allen Messungen gleichen Belastung ein charakteristisches Verhalten. Sie springen im Vergleich zum gesunden Knie nach vorn bzw. hinten. Der Arzt spricht von einer Schubladenbewegung, wir wollen einfacher von einer "Antwort" sprechen. Herr Stedtfeld und seine Mitarbeiter haben nun eine große Zahl von Knien untersucht. Es wurden Versuchsreihen betrachtet, in denen systematisch mehr und mehr verschiedene Bänder des Kniegelenkes durchtrennt waren und durch Röntgenaufnahmen die "Schubladengrößen" ermittelt wurden.
Es entstanden mehrere tausend Daten und es galt, einen Überblick über dieses Datenmaterial zu gewinnen, um daraus Informationen für die Diagnose Unfallverletzter zu bekommen.
Mein Vorschlag war, über der durch die beiden Koordinaten "Beugungswinkel des Knies und Drehwinkel des Fußes gegen die normale Vorwärtsrichtung (Verdrillung)" bestimmten Ebene als Funktion die Größe der Antworten abzutragen und diese Flächen graphisch darzustellen. Die Durchführung dieser Aufgabe übernahm mein Mitarbeiter, Herr Stenzel, am Institut für Angewandte Mathmatik der Universität Bonn. Die Resultate sind für die einzelnen Verletzungstypen charakteristische Flächen, welche dem Arzt eindrucksvoll vermitteln, was für Verletzungen vorliegen. Sie helfen ihm also wesentlich bei der Diagnose. (Vgl. Stedtfeld,...,[3])
Natürlich liegt bei den Antworten eine nicht unerhebliche Streuung vor und außerdem muß man für die späteren praktischen Anwendungen damit rechnen, daß statt der reinen Experimente mit intakten oder zertrennten Bändern in dem gegebenen Einzelfall einige Bänder nur gezerrt, andere zerrissen sind.
Aus diesem Grund erscheint es zweckmäßig, die Antworten nicht zahlenmäßig genau zu erfassen und auszuwerten, sondern nur zwischen geringen, mittleren und starken Antworten (durch 0,1,2 charakterisiert) zu unterscheiden. Gesucht war nun ein einfacher und für den Patienten schonender Weg, ein "Kriterium", zur Stellung der Diagnose, d.h. etwa die Durchführung einiger weniger Einstellungen von Beugungen des Knies und Drehungen des Fußes und Ermittlung zugeordneter Antworten, aus denen man die Art der Verletzung erschließen kann. Ein solches Kriterium muß aus den empirischen Werten (Messungen) gewonnen werden und bei der Vielzahl der Daten wird man sich dabei natürlich eines Computers bedienen.

Formulierung des mathematischen Modells

Die Verletzungstypen (wie etwa: ein Kreuzband zerrissen, ein Seitenband zerrissen, ein Kreuzband und ein Seitenband zerrissen,...) mögen mit den Indizes $i = 1, \ldots, l$, die verschiedenen Einstellungen (Tests) des Knies und Fußes mit $k = 1, \ldots, m$ durchnumeriert sein.

Zunächst kann man die gemessenen (und für gleichartige Verletzungen gemittelten) Daten zu einer Matrix $((a_{ik}))$ vereinigen, wobei a_{ik} die Antwort eines Knies beim Verletzungstyp i in der Stellung k ist.

Aufgrund der medizinischen Befunde und ihrer Auswertungen kennt man also zu jedem i die zugehörige Zeile a_{ik}, $k = 1, \ldots, m$.

Die Diagnose besteht nun gerade darin, aus einigen und zwar möglichst wenigen Antworten a_{ik} für Stellungen zum Index $k = k_1, k_2, \ldots$ auf den Zeilenindex i, d.h. den Typ der Knieverletzung zu schließen.

Unter der Annahme, daß man, wie wir es oben taten, nur 3 mögliche Werte als Antworten zuläßt, kann man offenbar mit einer Einstellung, d.h. der k_1-Spalte höchstens 3 Typen unterscheiden. Nimmt man 2 Spalten, so sind es maximal 3^2 Typen usw.

Im konkret vorliegenden Beispiel traten 19 Knieeinstellungen und 21 Verletzungstypen auf. Demnach wären wenigstens 3 Einstellungen k_1, k_2, k_3 erforderlich. Ob man mit 3 Einstellungen tatsächlich zu einer eindeutigen Diagnose gelangen kann, läßt sich mit dem nachfolgenden Algorithmus entscheiden. Da wir verhältnismäßig kleine Zahlenmengen zu untersuchen haben, spielt dabei die Frage der Komplexität keine Rolle, vgl. hierzu etwa Comer [1, 2], sondern es geht um eine konkrete Lösung des Problems.

Algorithmus zur Bestimmung der Minimalanzahl notwendiger Tests.

Vorgelegt sei die Matrix $((a_{ik}))$, i indiziert die verschiedenen Typen, $i = 1, \ldots, l$ und k die verschiedenen Einstellungen (Tests), $k = 1, \ldots, m$. Gibt es bis zu b Antwortmöglichkeiten, d.h. $a_{ik} \in \{0, \ldots, b-1\}$, so braucht man also wenigstens n Tests, wobei $b^{n-1} < l \leq b^n$ sei.

Wir prüfen, ob es wenigstens ein n–Tupel von Tests, charakterisiert durch $(k_1, \ldots, k_n)$ gibt, welches die eindeutige Identifizierung eines jeden Typs gestattet.

Es muß – soll die Aufgabe eine Lösung haben – einen Test in dem n-Tupel geben, der zwischen der 1. und 2. Zeile zu unterscheiden gestattet. Wir gehen also alle Indizes $k = 1, \ldots, m$ durch und prüfen, für welche $k_1 = k$ gilt $a_{1k} \neq a_{2k}$. Diese k_1 speichern wir als mögliche Elemente für ein Lösungs-n-tupel. Ist diese

Menge leer, gibt es keine Lösung.

Entweder es gilt auch noch $a_{1k} \neq a_{3k}$, $a_{2k} \neq a_{3k}$, dann trennt der k-te Test auch schon die ersten 3 Zeilen, oder es gilt $a_{3k_1} = a_{1k_1}$ bzw. a_{2k_1}. In diesem Fall müssen wir nach einem weiteren Test $k_2 \neq k_1$ suchen, so daß a_{3k_2} die Unterscheidung gestattet, d.h. es muß gelten $a_{3k_1} = a_{ik_1} \Rightarrow a_{3k_2} \neq a_{ik_2}$.

Wieder bricht das Verfahren ab, wenn mehr als 2 Zeilen vorhanden sind und kein Paar (k_1, k_2) existiert, welches diese 3 Zeilen zu trennen gestattet.

Auf diese Weise fährt man fort, indem man sukzessive eine Zeile nach der anderen hinzunimmt.

Man hat dann zu prüfen, ob die neue Zeile (mit Index j) über den ausgewählten Tests $(k_1, \ldots, k_s)$ mit einer früheren Zeile übereinstimmt:

$$a_{\nu k_\mu} - a_{jk_\mu} = 0 \quad \forall \mu = 1, \ldots, s$$

und muß einen weiteren Test k_{s+1} hinzuziehen um zu trennen, wenn es ein solches $\nu \in \{1, \ldots, j-1\}$ gibt.

Gibt es kein solches ν, so gestatten bereits $(k_1, \ldots, k_s)$, die ersten j Zeilen, d.h. Typen zu unterscheiden.

Bei der Frage der Lösbarkeit kommt es nun nur auf die Ermittlung einer Lösung an, und man wird etwa von den ersten möglichen k_1 ausgehend ein $k_2, \ldots$ bestimmen, bis entweder eine Lösung gefunden ist oder man feststellt, daß man mit diesen $k_1, \ldots$ nicht zu einer Lösung kommen kann. Man stößt also bei diesem Suchbaum so schnell wie möglich in die Tiefe.

Natürlich kann man in der skizzierten Weise auch alle Lösungen ermitteln, wenn man alle Zweige des Suchbaums abfährt, bis man jeweils entweder eine Lösung gefunden hat oder die Unlösbarkeit mit einer gewissen Kombination feststellen mußte.

Wir betrachten ein einfaches fiktives Beispiel.

Sei

$$((a_{ik})) = \begin{pmatrix} 0 & 1 & 2 & 1 & 2 & 0 \\ 0 & 1 & 1 & 2 & 1 & 0 \\ 0 & 1 & 0 & 1 & 0 & 1 \\ 0 & 2 & 2 & 2 & 2 & 0 \\ 1 & 1 & 1 & 1 & 1 & 2 \\ 2 & 2 & 1 & 1 & 2 & 1 \end{pmatrix}.$$

Kann man hier mit zwei Tests trennen? Es ist ja $b = 6 < 3^2$.

Hier kommen zur Trennung der ersten zwei Zeilen nur die Indizes

$$k_1 = 3, 4, 5$$

in Frage.

$k_1 = 3$ und 5 gestattet auch die Trennung der ersten 3 Zeilen.

Will man davon auch noch die 4. Zeile unterscheiden, so kann man mit den beiden Tests mit den Indizes

$$k_2 = 2, 4$$

kombinieren.

Die 5. und 2. Zeile werden jedoch von (2,3) nicht getrennt, die 5. und 6. Zeile nicht von (3,4). Damit fällt $k_1 = 3$ aus.

Man bemerkt jedoch sofort, daß diese Trennung durch Hinzunahme eines weiteren Tests etwa mit $k = 5$ möglich ist.

Also muß man einen neuen Ansatz etwa mit

$$k_1 = 4 \qquad \text{machen.}$$

Schon zur Trennung der 3. Zeile muß man einen weiteren Test heranziehen. Mögliche Werte sind

$$k_2 = (3), 5, 6.$$

Die geklammerten Werte wurden bereits oben bearbeitet.

Man sieht sofort, daß das Testpaar (4,5) die 4. und 5. Zeile nicht trennt, daß (4,6) die Zeile 1 und 4 nicht unterscheidet.

Es bleibt also nur noch

$$k_1 = 5 \qquad \text{zu testen.}$$

Hier werden zwar die ersten 3 Zeilen getrennt, für die 4. Zeile muß man aber einen weiteren Test heranziehen:

$$k_2 = 2, (4) \qquad \text{käme in Frage.}$$

Davon wurde das Paar (4,5) bereits ausgeschlossen. Von (2,5) werden jedoch die 2. und 5. Zeile nicht getrennt.

Also kann man nicht mit zwei Tests die Zeilen dieser Matrix identifizieren, sondern muß mindestens 3 Tests machen.

Zur Aufzählung aller Lösungen bei s Tests soll ein zweiter Algorithmus skizziert werden. Er wird ebenfalls an dem obigen Beispiel verdeutlicht. Über die Ergebnisse für die medizinischen Originaldaten wird an anderer Stelle berichtet.

Algorithmus zur Bestimmung aller identifizierenden s–Tupel von Tests bei Vorgabe der Anzahl s .

Es sei also s gegeben. Es gebe m Tests, sei $\mathsf{K} = \{1,\ldots,m\}, \mathsf{L} = \{1\ldots,l\}$. Es geht dann darum, bei Vorgabe einer Matrix mit den Antworten a_{ik} alle s-Tupel $(k_1,\ldots,k_s)$ zu ermitteln, durch welche eindeutig aus den Meßwerten $m_1,\ldots,m_s$ mit $m_j \in \{0,1,\ldots,b-1\} =: \mathsf{B}$ durch die Gleichungen

$$a_{ik_\nu} = m_\nu$$

ein Index i ausgezeichnet ist. Dabei braucht nicht für jeden Vektor von Meßwerten eine Lösung zu existieren.

Der Algorithmus wird nicht durchformalisiert, sondern nur für das oben bereits verwendete Beispiel beschrieben. Dort zeigte sich, daß man 3 Tests zur Identifikation machen muß. Die Verallgemeinerung ist klar und wird dem Leser überlassen.

Für $s = 3$ Tests sollen alle Lösungen aufgestellt werden. Im allgemeinen Falle wird man höchstens $l \leq b^3 = 3^3$ Typen unterscheiden können und zunächst jeden Index als k_1 in Betracht ziehen, der eine Aufteilung der Typen in 3 Klassen bewirkt, von denen jede höchstens $b^2 = 3^2$ Elemente enthält, so daß durch 2 weitere Tests die Möglichkeit einer Trennung besteht. Setzt man

$$l_1(i,\nu,k) = \begin{cases} 1 & \text{falls } a_{ik} = \nu \\ 0 & \text{sonst} \end{cases} \quad \text{für } \nu \in \mathsf{B},$$

so charakterisiert $l_1(i,\ldots)$ bei festem ν und k gerade diejenigen Typen, welche die Antwort ν bei der Einstellung k geben, also zur Klasse

$$M_1(\nu,k) := \{\, i \mid a_{ik} = \nu, \, i \in \mathsf{B} \,\}$$

gehören. Die Anzahl der Elemente von $M_1(\nu,k)$ ist also durch

$$|M_1(\nu,k)| = \sum_i l_i(i,\nu,k)$$

gegeben. In der Menge K_1 sollen nun, mit einem Index j durchnumeriert, diejenigen $k = k_j$ gesammelt sein, für die für jedes zulässige ν die Abschätzung

$$|M_1(\nu,k_j)| \leq 3^2$$

gilt. (In unserem Modellfall ist diese Bedingung wegen $l = 6$ trivialerweise erfüllt.) Die Bestimmung der l_1 und der Menge K_1 erfordert $\mathcal{O}(lm)$ Operationen.

Nur Elemente von K_1 kommen als weitere Kandidaten k_1 für die genannten Tripel in Frage. Als Antwort auf zwei Einstellungen (k_1, k_2) wird man zwei Werte $(\nu, \mu) = (a_{ik_1}, a_{ik_2})$ erhalten und auf die Frage, für welche Typen i diese Antworten auftreten, gibt die Menge

$$M_2(\nu, \mu; k_1, k_2) = \{\ i\ |\ l_2(i, \nu, \mu; k_1, k_2) := l_1(i, \nu, k_1) \cdot l_1(i, \mu, k_2) = 1, i \in B\ \}$$

Antwort.

Da nur noch eine weitere Messung folgen soll, kann man nur solche (k_1, k_2) gebrauchen, für die für jedes zulässige Paar ν, μ die Abschätzung

$$|m_2(\nu, \mu; k_1, k_2)| = \sum_i l_2(i; \nu, \mu; k_1.k_2) \leq 3, \quad k_1 \in K_1, \quad k_2 \in K,$$

gilt.

Man faßt alle Tupel (k_1, k_2), für welche diese Bedingungen erfüllt sind, zu $\underline{K_2}$ zusammen.

Analog zu diesem Schritt für die Aufstellung von K_2 verfährt man bei der Untersuchung der Tripel. Man kombiniert $(k_1, k_2) \in K_2$ mit $k_3 \in K$, wobei es wegen der Vertauschbarkeit genügt, nur die $k_3 > k_2$ zu betrachten.

Setzt man

$$l_3(i; \nu, \mu, \rho; k_1, k_2, k_3) = l_2(i; \nu, \mu; k_1, k_2) \cdot l_1(i; \rho; k_3)$$

und

$$M_3(\nu, \mu, \rho; k_1, k_2, k_3) = \{\ i\ |\ l_3(i; \nu, \mu, \rho; k_1, k_2, k_3) = 1,\ i \in B\ \}.$$

so ist

$$|M_3(\ldots)| = \sum_i l_3(i, \ldots)$$

und man wird definieren

$$K_3 = \{\ (k_1, k_2, k_3)\ |\ \forall \nu, \mu, \rho : |M_3(\nu, \mu, \rho; k_1, k_2, k_3)| \leq 1\ \}$$

Bisher wurden notwendige Bedingungen berücksichtigt.

Ein Element (k_1, k_2, k_3) von K_3 löst nun das Diagnoseproblem, wenn alle Verletzungstypen als Antworten vorkommen, d.h.

$$(*) \qquad \bigcup_{\nu, \mu, \rho} M_3(\nu, \mu, \rho; k_1, k_2, k_3) = \{1, \ldots, l\} = B$$

gilt.

Wir bemerken, daß bei festem $(k_1, k_2, k_3) \in \mathsf{K}_3$ die Zuordnung

$$(\nu, \mu, \rho) \to i$$

umkehrbar eindeutig ist.

Denn $i \to (\nu, \mu, \rho)$ ist nach Definition der Matrix $((a_{ik}))$ eindeutig und die Umkehrung steckt in der Definition von K_3.

Es genügt also, statt (*) die Forderung

$$\sum_{\nu, \mu, \rho} \sum_i l_3(i; \nu, \mu, \rho; k_1, k_2, k_3) = l$$

für jedes Tripel $(k_1, k_2, k_3) \in \mathsf{K}_3$ zu erfüllen, einfacher ausgedrückt, in jeder Spalte der von den Elementen $l_3(i, \nu \cdot 3^2 + \mu \cdot 3 + \rho)$, Zeilenindex $i \in \mathsf{B}$, $\nu, \mu, \rho \in \mathsf{B}$, gebildeten Matrix muß genau eine 1 stehen, wobei $l_3(i, \nu \cdot 9 + \mu \cdot 3 + \rho) :=$ $l_3(i; \nu, \mu, \rho; k_1, k_2, k_3)$. Durch diese Matrix wird in gewisser Weise die Matrix $((a_{ik}))$ invertiert: $(\nu, \mu, \rho) \to i$.

Im vorliegenden Beispiel fiktiver medizinischer Daten ist diese Umkehrung möglich, wie die folgenden Überlegungen zeigen.

Zunächst kann man jedem Test bzw. jeder Menge von Tests, (gekennzeichnet durch ihre Indizes und in runde Klammern gesetzt) die Mengen zuordnen, in welche die Verletzungstypen gegliedert werden (diese Mengen in geschweifte Klammern gesetzt). Diese Organisiation ist sparsamer im Speicherplatzverbrauch als das formale Befolgen der vorangehenden Konstruktion.

Im vorliegenden Beispiel ist die Zuordnung unmittelbar abzulesen:

$$(1) \quad \to \quad \{1, 2, 3, 4\} \ \{5\} \ \{6\}$$

Man braucht nun die einelementigen Mengen nicht weiter zu betrachten, denn sie charakterisieren die Typen 5 und 6 bereits eindeutig, sondern kann sich darauf beschränken, die mehrgliedrigen Mengen zu differenzieren.

$$(1, 2) \quad \to \quad \{1, 2, 3\} \ \{4\} \ \{5\} \ \{6\} \qquad \text{und} \qquad (1, 2, 3)$$
$$\text{ebenso wie} \quad (1, 2, 5)$$

gestatten die Trennung aller Typen, da (3) ebenso wie (5) die 3 Typen 1,2,3 trennt.

Analog sieht man

$$(1,3) \rightarrow \{1,4\} \ \{2\} \ \{3\} \ \{5\} \ \{6\} \quad \text{und}$$

$$(1,4) \rightarrow \{1,3\} \ \{2,4\} \dots \{6\} \quad \Rightarrow$$

$$(1,5) \rightarrow \{1,4\} \ \{2\} \dots \{6\} \quad \Rightarrow$$

$(1,3,4)$	trennt
$(1,4,5)$	
______	keine Lösung,

da $\{1,4\}$ von (6) nicht getrennt wird.

$$(2) \rightarrow \{1,2,3,5\} \ \{4,6\}, \quad \text{dies führt zu}$$

$$(2,3) \rightarrow \{1\} \ \{2,5\} \dots \{6\} \quad \Rightarrow \quad (2,3,4), (2,3,5)$$

$$(2,4) \rightarrow \{1,3,5\} \dots \{6\} \quad \Rightarrow \quad (2,4,5), (2,4,6)$$

$$(2,5) \rightarrow \{2,5\} \ \{4,6\} \quad \Rightarrow \quad (2,5,6)$$

$$(3) \rightarrow \{1,4\} \ \{2,5,6\}$$

$$(3,4) \rightarrow \{5,6\} \quad \Rightarrow \quad (3,4,5), (3,4,6)$$

$$(3,5) \rightarrow \{1,4\} \ \{2,5\} \quad \Rightarrow \quad \text{______ keine Lösung}$$

$$(4) \rightarrow \{1,3,5,6\} \ \{2,4\}$$

$$(4,5) \rightarrow \{1,6\} \quad \Rightarrow \quad (4,5,6)$$

Man bekommt also in diesem Beispiel 12 Dreierkombinationen von Tests, die eindeutig auf den vorliegenden Verletzungstyp zu schließen gestattet.

Bei kleinen Zahlen s von Tests stellt die Suche offenbar kein großes Problem dar. Asymptotisch kann man jedoch zeigen, daß der Aufwand (Speicher und Zeit) zu einem stärkeren als polynomialen Wachstum führt, so daß bei Informatikern die Lösung des Problems als schwierig gilt, vgl. Comer et al [1, 2].

Hier sei nur zum Abschluß angemerkt, daß in natürlicher Weise die Verallgemeinerung dieses Identifikationsprozesses auf andere Matrizen von Typen von Verletzungen und Messungen möglich ist. Gibt es mehrere Lösungen, wird der Arzt eine medizinisch geeignete auswählen, die den Patienten möglichst wenig belastet. Er wird beispielsweise extreme Fußdrehungen und Beugungen des Knies vermeiden. Außerdem werden durch einen Test, in der Praxis eine Röntgenaufnahme, mehrere Daten gekoppelt, also eigentlich ein Vektor von Meßwerten, erhoben. Bei der von uns vorgenommenen Diskretisierung ist dies jedoch zum obigen skalaren Fall äquivalent, denn man kann einfach alle endlich vielen Fälle durchnumerieren, d.h. b hinreichend groß wählen.

Natürlich liegen in der klinischen Praxis nicht die reinen Fälle vor, die wir als Ausgangsmaterial für unsere Untersuchungen verwenden können. In der praktischen Anwendung muß man also noch eine gewisse Streuung der Antworten berücksichtigen, dieser Tatsache wurde hier durch eine Klassifizierung Rechnung zu tragen versucht. Herr Stenzel führt dazu ein gewisses Distanzmaß ein. Für eine Diskussion dieser Probleme sei auf eine geplante Veröffentlichung mit den medizinischen Ergebnissen verwiesen.

Herrn Professor Dr.Ottmann, Karlsruhe, danke ich für den Hinweis auf die Arbeiten von Comer et al.

<u>Literatur</u>

[1] Comer, D. and Sethi, R.: The Complexity of Trie Index Construction.
   Journal of the ACM 24 (1977), 428-440
[2] Comer, D.: Analysis of a Heuristic for Full Trie Minimization.
   ACM Transactions on Database Systems 6 (1981), 513-537.
[3] Stedtfeld, H-W., Strobel, M., Stenzel, H.,: Beitrag zur Frage der
   Postero-medialen Instabilität des Kniegelenkes.
   Eingereicht zur Publikation

Anschrift des Verfassers:
Professor Dr. Helmut Werner,
Institut für angewandte Mathematik der Universität Bonn
Wegelerstr.6, 5300 Bonn 1, Deutschland